Rational Animals

# Rational Animals

## The Teleological Roots of Intentionality

MARK OKRENT

OHIO UNIVERSITY PRESS    ATHENS

Ohio University Press, Athens, Ohio 45701
www.ohio.edu/oupress


Printed in the United States of America


Ohio University Press books are printed on acid-free paper ∞ ™

14 13 12 11 10 09 08 07     5 4 3 2 1

*Library of Congress Cataloging-in-Publication Data*

Okrent, Mark, 1947–
Rational animals : the teleological roots of intentionality / Mark Okrent.
p. cm. — (Series in Continental thought ; 34)
Includes bibliographical references and index.
ISBN-13: 978-0-8214-1743-0 (cloth : alk. paper)
ISBN-10: 0-8214-1743-6 (cloth : alk. paper)
ISBN-13: 978-0-8214-1744-7 (pbk. : alk. paper)
ISBN-10: 0-8214-1744-4 (pbk. : alk. paper)
1. Intentionality (Philosophy) I. Title.
B105.I56O37 2007
128'.3—dc22

2007011356

*To my children, Nick, Valerie, Jon, and Anna*

*And to my life partner, Charlotte Witt, who supported me through the difficult development of this book*

# CONTENTS


Preface ix

Style and Substance ix
The Structure of the Book xv

Introduction. Ethological Anecdotes, Philosophical Pronouncements 1

Chapter 1. What Is Intentionality? 12

1. Three Central Issues Concerning Intentionality 12
2. The Pragmatic Approach to Intentionality 17
3. Representationalism, Functionalism, and Pragmatism 22

Chapter 2. Goals without Intentions 28

1. The Field Teleologist's Problem 28
2. Goal-Directedness: Description and Explanation 31
3. Wright's Analysis of Goals 40

Chapter 3. The Descriptive Content of Goal Attributions 51

1. The Scope of the Judgment of Appropriateness 51
2. Goals as Strongly Holistic Properties 54
3. Living Agents and Teleology 60

Chapter 4. The Explanatory Power of Goal Attributions 75

1. How Goal-Directed Attributions Explain 75
2. Goals and Functions: Types and Individuals 83
3. Types and Goals: Populations and Evolution 89

Chapter 5. Instrumental Rationality 104

1. The Teleological Stance and the Intentional Stance 105
2. Reasons 108
3. Nonrational Behavior: The Lessons of the *Sphex* Wasp 112
4. Instrumentally Rational Behavior: The Lessons of the Plover 118
5. Goals and Reasons 122
6. The Norms of Instrumental Rationality 127

Chapter 6. Rational Explanation 134

1. Justification and Explanation 134
2. Inference and Explanation 141
3. Goals and Desires 148
4. The Motivation of Rational Behavior 153
5. Rational Explanation and Intentional Content 161
6. Summary 163

Chapter 7. Practical Reason and Definite Content 166

1. The Limitations of Merely Instrumental Reason 166
2. Opacity and Vague Contents 168
3. Reflection and Practical Rationality 176
4. The Evolution of Practical Rationality 183
5. Practical Reason and Social Roles 189
6. Tools, Language, and Definite Content 196

Notes 205
Bibliography 215
Index 219

# PREFACE

## 1. STYLE AND SUBSTANCE

Intentionality, aboutness, or directedness toward, is a mystery.

On the one hand, that some entities (in the broadest sense) are about something seems obvious to us minded creatures. Our primary ways of understanding what we are and what we do depend on our understanding ourselves as having states that are about or directed toward things other than ourselves. It is even the case that we understand ourselves as having some states that are reflexively about ourselves or our other intentional states. Our perceptions are perceptions of things. Our beliefs characterize entities as this or that. Our hopes, fears, and wants are attitudes toward possible ways in which things might be. What we say has "meaning" and "refers" to entities that we are not. When we act, what we do is directed toward or is in order to achieve some goal, and that goal is often some possible way in which something other than ourselves might be. We are capable of having attitudes, ranging from pride to loathing to hope and joy, about our own mental states.

On the other hand, we really don't have a very good grasp on what intentionality consists of or how it is possible. We moderns tend to understand the natural world in terms of attitudes that derive from the scientific revolution and the Enlightenment. And those attitudes, which emphasize that the universe of entities is coextensive with the realm of nature and that nature is determined by laws defined over physical characteristics, appear to leave no room for directedness or aboutness. From the standpoint of the physical, intentionality appears just plain weird. For example, intentionality seems to involve relatedness. But real, natural relations demand that all the things that are related must exist. If Valerie is taller than Anna, both Anna and Valerie must exist. But intentional relations appear to violate this rule. I can think about my nonexistent sixth child, talk about unicorns, and—crucially—desire, dread, hope for, and attempt to achieve possibilities that might never be or become actual. Similarly, it seems central to intentional states that, as such, they can be normatively evaluated and that those evaluations are objective. Beliefs ought to be true, and whether or not they are true is objective. Inferences ought to be valid, and whether they are valid is an objective fact about those inferences. Actions are, objectively, failures or successes. But we take it that we can give a complete description of the natural world (and for us that means the real world) that makes no reference to objective values at all. So how can there *be* a realm of the intentional at all? And, if there is, how can we understand what it is to be intentional?

To a very large extent, the mystery surrounding intentionality was the defining issue of twentieth-century philosophy. This was as true for the analytic tradition of philosophical thought that began with Frege and Russell as it was for the phenomenological tradition that began with Husserl, even if the centrality of intentionality as a problem was not always emphasized as explicitly in the first tradition as it was in the second. This fact is evidenced both by the centrality of philosophy of language to analytic philosophy in the first two-thirds of the century and by the rising importance of philosophy of mind in the remainder of the century. Both language and mind are, essentially, intentional phenomena, and analytic considerations of these phenomena center on their intentional aspects, such as reference, meaning, mental content, and failure, within intentional contexts, of the substitution of coreferring expressions to preserve the truth-value of sentences.

Although analytic attempts to understand intentionality canvassed a wide range of alternatives throughout the course of the century (including behaviorism, type-identity theory, a kind of functionalism rooted in an evolutionary understanding of biological functions, and even some kinds of dualism), by the beginning of the twenty-first century informed opinion came to center on causal-role functionalism as the preferred model. Although this style of functionalism has considerable difficulty comprehending consciousness, a consensus has emerged that intentionality itself can be understood in terms of causal roles. This consensus has emerged even though such an account of intentional states faces several outstanding difficulties.

From the perspective that I develop in this book, causal-role functionalism has some distinct virtues. In particular, it is a view of the intentional mind that emphasizes the way in which intentional mental states provide an explanatory link between the manner in which a system is affected by its environment in perception and the manner in which that system responds to that environment in action. Nevertheless, for reasons that will become clear in the course of this book, it seems to me that this style of functionalism is simply hopeless as a general account of intentionality. Briefly, these reasons turn on the fact that causal-role functionalism does not have the resources to account for the normativity that is central to intentionality. This lack of resources arises from two related factors. First, causal-role functionalism assumes that action explanation is essentially causal. Second, causal-role functionalism has an especially sparse notion of what an act is, a notion that leaves no room for the fact that acts themselves can be teleologically directed toward goals, independent of their causal relations to intentional mental states.

In this book, I present an alternative understanding of intentionality that focuses on the relation between the intentionality of mental states and the teleological goal-directedness of action. I argue that the primary form of intentionality is action and that actions are intentional in that they are directed toward goals. That is, I suggest that, in a specific sense to be developed, all intentionality depends on the goal-directedness of activity. Intentionality is rooted in teleology; the aboutness of mental life is rooted in the goal-directedness of active life.

I make no claim for the originality of this thesis. It appeared several times in the course of twentieth-century reflections on intentionality, most prominently in the work of Dewey, Merleau-Ponty, the early Heidegger, and the late Wittgenstein. I myself borrowed the outlines of this thesis from *Being and Time*. In my 1988 book, *Heidegger's Pragmatism*,[1] I suggested that for the early Heidegger, no agent could have intentional states without possessing what Heidegger called "understanding" and that no agent could possess understanding unless it was capable of acting in order to achieve some end or goal. That is, the central thesis of this book is also, on my reading, one of the central claims of *Being and Time*. There can be no question that Heidegger, and not Okrent, is the source of this position. Beyond this reliance on Heidegger for the central thesis of the book, many of the details of my understanding of the teleological character of action and the relation between intentionality and teleology come directly and indirectly from my reading of Merleau-Ponty.

This dependence on Continental sources might lead one to expect that this book would be about one or another Continental figure, or at least written in a recognizably Continental style. Nevertheless, this current work is not about Heidegger or Merleau-Ponty or any other historical figure. It is not even written in a recognizably Heideggerean or Continental style. In addition, the contemporary figures with which it engages come from the analytic rather than the Continental tradition. Given that I acknowledge and emphasize what is obvious, that the central thesis and inspiration for this work derive from Heidegger, these facts deserve explanation and justification.

Two central considerations account for the distinctive style and tone of *Rational Animals*. One is substantive; the other is strategic. To start with the substantive consideration, in certain crucial respects this book is not a work in Heideggerean philosophy. Although the position that I develop in this book derives directly from Heidegger's work, it deviates from Heidegger's position in two fundamental respects. I claim that while goal-directed action is a necessary condition on all intentionality, as Heidegger taught, the distinctively human style of action that implicitly involves understanding in Heidegger's sense is, in the position I develop here, not a necessary condition on goal-directed action as such. Creatures that are not Dasein act in order to achieve ends. Second, in my view, beings such as ourselves, Dasein that are capable of a linguistic being-in-the-world and an understanding of being, are intelligible only as modifications of simpler forms that act in order to achieve ends of their own—and sometimes even act for reasons of their own—but are incapable of the kind of intentions directed toward themselves that we routinely display. That is, we are a distinctive kind of rational animal. This is a manifestly anti-Heideggerean view.

The belief that the proper home of intentionality is exclusively with beings like ourselves, who are capable of an understanding of their own being as well as an understanding of the being of other entities, is not unique to Heidegger. In one way or another, the position has been common to both the analytic and the Continental wings of the tradition that arose in Kant. Davidson, Sellars, and Brandom are as committed to this doctrine as is the early Heidegger or Gadamer. All of these thinkers share the belief that the necessary conditions on the kind of linguistically articulated intentionality that is the human condition are at the same time the necessary conditions on intentionality as such. From the standpoint developed in this book, this prioritizing of human, Daseinish intentionality is both an unsupported dogma and a position that makes it impossible to understand how beings that are capable of linguistic intentionality are rooted in their animal natures or how properly to comprehend the relation between intentional thought and purposeful action.

From the perspective I present in this book, Heidegger's early work on the conditions on intentionality involved two seminal insights, above all others. The first was the recognition that the intentionality of thought was dependent on the directedness of action. This, as outlined above, is the central insight that has motivated the overall position developed in *Rational Animals*. The second insight was articulated in Heidegger's profound analysis of the being of Dasein. What Heidegger achieved with the analysis of Dasein's being was not what he had intended, however—the articulation of the necessary conditions on all intentionality. What this analysis actually accomplished was a remarkably profound articulation of the way in which Dasein is distinctively different from all other beings that act in order to achieve goals and that act for reasons. For this reason, chapter 7 of this book, which attempts to understand how human linguistic intentionality and goal-directedness are related to and distinct from mere animal intentionality and goal-directedness, is heavily and explicitly dependent on my interpretation of Heidegger. The earlier portions of *Rational Animals* place this discussion in a fundamentally anti-Heideggerean context, however, even if it is a context that was inspired by my reading of Heidegger. In my view, Dasein is and must be a kind of rational animal if it is to be Dasein at all. This is not a position that Heidegger could ever accept. Because of these significant deviations from Heideggerean orthodoxy, it would have been at best misleading to present these views as an interpretation or development of Heidegger's thought, even though they derive from that thought. For that reason, I have not framed my discussion in Heideggerean terms.

Beyond these substantive considerations, there was a set of strategic factors that influenced the style in which these investigations were carried out. It has always seemed to me that the analytic and Continental traditions of twentieth-century philosophical thought really formed a single discursive space but that this continuity was masked by merely stylistic differences of tone and contingent differences in the trails of citations. To put the point differently, it seems to me that the major differences between these two ways of doing philosophy concern merely the fact that Continental and analytic authors read differently and cite different predecessors in different ways. These superficial differences, however, have rendered it remarkably difficult for work in one of these styles to get a careful and sympathetic reading by thinkers used to working in the other style.

This dialectical situation is especially unfortunate for two reasons. First, as mentioned above, both traditions have been centrally concerned with the same set of issues regarding intentionality. It is not as if the problems with which the traditions are concerned are in any way incommensurable. But second, there is sufficient overlap, as well as difference, between some of the positions developed by various groups in the two traditions to make a dialogue especially fruitful. In particular, the philosophers working in the analytic schools that derive from Quine and Sellars, on the one hand, and those philosophers whose work on intentionality is informed by Heidegger and Merleau-Ponty, on the other, have a great deal to learn from one another. So anything that any of us who work in these traditions can do to facilitate a dialogue between these respective camps will bring, I believe, distinct benefits to our understanding of the puzzles surrounding intentionality.

In 1988, when *Heidegger's Pragmatism* was published, there was a strong prejudice in analytic circles against Continental philosophy in general, a bias that made it difficult for a philosopher like me, working on a figure like Heidegger, to get a fair and open hearing. (The fact that there was an equal and opposite prejudice against analytic philosophy in Continental circles also didn't help matters much.) Nevertheless, the importance I attached to dialogue with analytic philosophers gave me a strong motive to articulate the parallels I found between Heidegger and Husserl, on the one hand, and Quine, Davidson, and Sellars, on the other, in such a way that thinkers familiar with those analytic figures could see the points of connection.[2] Fortunately, the dialectical situation has changed significantly since then. Although there are certainly still many analytic philosophers who have an antipathy toward Continental philosophy as such, a critical mass of thinkers in the analytic tradition has a far more receptive attitude toward work with Continental roots than used to be the case. Nevertheless, openness toward philosophical views that derive from Continental sources has not bred a familiarity with and fluency in the Continental style of doing philosophy. That is, analytic philosophers might be open to and interested in work with Continental roots, but many of them still find it difficult to read and understand Continental philosophy when it is carried out in its own native idiom.

It follows from this constellation of factors that any author influenced by the Continental tradition who wishes his work to be received with understanding by an analytic audience as well as a Continental one had best present his ideas in a style that is accessible to the widest possible public rather than insisting on stylistic purity. This consideration has combined with the fact that many of the theses in this book depart significantly from their Heideggerean origins to motivate the form in which the ideas are presented.

## 2. THE STRUCTURE OF THE BOOK

In this book I argue that the intentional content of mental states derives from the roles those states play in the rational explanation of goal-directed action. It long has been recognized that there is a tight connection between the intentional contents of an agent's mental states and an agent's rational behavior. After all, under normal circumstances, the only evidence we have for what an agent believes and desires, or even whether an agent has any beliefs or desires at all, is what that agent does when it acts rationally. Indeed, the concepts of belief and desire, on the one hand, and of rational action, on the other, are so intimately connected that if one could understand either of them, one could also understand the other. From this standpoint, the only real question for a philosopher interested in rational action and the contents of mental states is which of these concepts one should use to help understand the other.

Perhaps because in standard cases of human action we explain the fact that an agent acts as it does by appealing to its beliefs and desires, the majority of twentieth-century philosophers attempted to understand what it is for an agent to act rationally by first understanding what it is for an agent to have beliefs and desires that could count as reasons for the agent's action. I argue that this procedure is exactly backward: to understand what it is for an agent to have beliefs and desires, one must first understand what it is for an agent's behavior to count as rational and then understand what it is for a state of an agent to be a belief or desire in terms of the potential roles of those states in the rational explanation of such rational behavior. There is an obvious problem with this strategy. To make it work, one must supply an account of what it is for an agent's behavior to be rational that is independent of and prior to an account of what it is for an agent to have beliefs and desires that could serve as reasons for that behavior. This problem is overcome by (1) emphasizing that rational action is a distinct species of goal-directed behavior, (2) developing a new account of what is involved in an organic agent's acting nonrationally to achieve a goal, and (3) articulating the features of instrumentally rational behavior that distinguish it as a species of goal-directed behavior. I argue that there are very good reasons for thinking that all animals act in a goal-directed fashion, while only the members of a small number of species act in an instrumentally rational fashion to accomplish their goals. I go on to specify the way in which the contents of mental states arise out of their roles in the kind of (rational) explanation that is necessary to account for an agent's rational action and to show how the type of practically rational behavior and the kind of highly definite intentional contents characteristic of human action and intentionality are related to and arise out of the merely instrumental rationality and vague mental contents characteristic of nonlinguistic animals.

The book is divided into seven chapters. In chapter 1 I distinguish three crucial puzzles that any account of intentionality must adequately treat. First, any such account must explain how states with intentional contents fit into a nature that can always be described accurately in physical terms. Second, any account of intentionality must adequately address the following puzzle regarding the logic of intentional states. If, on the one hand, the property of a state's being about or directed toward something is a relational property, as on the surface it appears to be, then how can a property involve a relation in which one of the relata need not exist? If, on the other hand, intentionality is an intrinsic property of a state, then how can such an intrinsic property contain an apparent reference to another state or entity? Third, and most important, any adequate account of intentionality must render intelligible the kind of intrinsic normativity that is involved in states with intentional contents. Insofar as beliefs, for example, are individuated by their contents, such states are essentially evaluable according to the norm of truth. But what fact about a belief makes it the case that, in itself, so to speak, it can be true or false and it *should* be true? I argue that neither causal-role functionalism nor any of the many varieties of representationalism can adequately account for the normativity associated with intentionality.

In chapters 2 through 4 I accomplish three tasks. First, drawing on the philosophical work of Larry Wright and Ruth Millikan as well as the ethological literature from biology, I argue that there are very good reasons to believe that the behavior of some animals is explicable by appealing to the goals of the behavior even though those animals do not act rationally. Second, I criticize Wright's specific account of what it is for an act to be for the sake of achieving some goal and Millikan's evolutionary account of what it is for some structure or behavior to have a function. Third, I develop a new account of the nonrational goal-directedness of behavior.

Drawing on a variety of ethological descriptions, in chapter 5 I establish criteria for distinguishing rational from nonrational goal-directed behavior. It is further argued that although no agent can count as acting rationally unless it counts as acting for goals, the kinds of goal-directed explanations that are adequate to explain and predict the nonrational goal-directed behavior of agents are not adequate to predict and explain in a detailed fashion the rational action of instrumentally rational agents. For that task, a different type of explanation is required, a rational explanation. In chapter 6 I articulate the structure of rational explanation and develop a theory of intentional content that derives the contents of beliefs and desires from their roles in the rational explanation of the instrumentally rational species of goal-directed action.

In chapter 7 I show how the kind of noninstrumental rational action that we human beings sometimes engage in both arises out of and is made possible by the kind of instrumentally rational behavior we share with other, nonlinguistic, animals. I further argue that this human kind of rationality allows for the possibility of articulate language and that such language use in turn makes it possible for humans to have intentional states with contents far more definite than those of the states of merely instrumentally rational animals.

*Rational Animals* thus brings together three lines of philosophical research that have important connections but that all too often have been conducted in strict isolation. These three areas concern (1) the proper way to understand the teleological descriptions and explanations that are endemic in ordinary life and biological writing, (2) the proper way to understand the implications of such biological teleological descriptions and explanations for issues in the philosophy of mind having to do with the character of rationality and intentional content, and (3) the implications of such naturalistic work on intentionality and rationality in the philosophy of mind for issues concerning ethics and practical reasoning. The overall strategy of *Rational Animals* shows exactly how and in what way these issues are related. While important work has been done in regard to teleological explanation by Robert Cummins, Peter Godfrey-Smith, and Philip Kitcher, as well as by Millikan, Wright, and many others, much of it has focused on functional explanation to the exclusion of an interest in goal-directed explanations. Because of this exclusive focus on functions, when philosophers of mind, such as Millikan and Daniel Dennett, have looked to teleology to provide illumination for our understanding of intentionality, they have concerned themselves exclusively with the possible biological functions of mental states. *Rational Animals* argues that this is a mistake and that the proper way to understand the connection between teleological and intentional discourse is through an understanding of the goals of behavior rather than through an understanding of the functions of mental states. It then becomes possible to show how nonreductionist naturalistic accounts of rationality and intentionality can help to illuminate the many puzzles regarding ethics and human practical reasoning that surround what Christine Korsgaard has labeled the question of normativity.

INTRODUCTION

# ETHOLOGICAL ANECDOTES, PHILOSOPHICAL PRONOUNCEMENTS

Ornithologists and ethologists have repeatedly observed the behavior of nesting plover when a large intruder . . . approaches a nest where a killdeer or piping plover is incubating its eggs. At a considerable distance, . . . the plover may stand up and walk slowly to a point a few meters from the nest. . . . The bird may then walk rapidly or fly in almost any direction except toward the nest. If a person approaches these birds while feeding or when they have no eggs or young, they fly away from the intruder to a safe distance. . . . The bird flutters slowly but conspicuously away from the nest, staying relatively close to the intruder. . . . If the bird is walking or running, its gait is different from normal locomotion and more noticeable. . . . Often the tail almost drags on the ground, and the wings are slightly extended, sometimes one more than another, strongly suggesting some weakness or injury. After running a few meters, the bird may flop about on the ground, extending one or both wings, as if injured. This is often called the "broken wing display," and it requires considerable effort for an observer to believe that the bird is really quite healthy. . . . Throughout most of this predator-distraction behavior, the bird watches the intruder. Typically it does not move in a straight line and stops from time to time. If the intruder approaches, the bird moves farther ahead. If not, the bird usually flies back closer to the intruder and repeats the behavior. . . . Typically the bird continues the injury simulation while leading the intruder some distance away from the nest or young. Finally, however, it flies away rapidly, usually in the same

> direction, then circling back to the general vicinity, though seldom to the exact spot, where the eggs or young are located.[1]

Donald Griffin, the ethologist who presents this description of the plover's behavior, concludes that the behavior constitutes evidence that the bird engages in episodes of conscious thought. It seems to me that Griffin overreaches here. What the plover displays, Griffin tells us, is "versatile adaptability of behavior to changing circumstances and challenges."[2] I will argue that such behavior provides evidence only for the proposition that the plover has beliefs and desires, not that the plover is conscious of having those beliefs and desires. To warrant such an attribution of consciousness, one needs a different sort of evidence.

What is philosophically interesting about Griffin's argument, however, is that the *evidence* he appeals to is the versatility and adaptiveness of the behavior. This adaptive behavior is evidence of intentional states because it is the explanandum that those states are needed to explain. Griffin's argument is as follows: The bird's behavior displays adaptive versatility. The best explanation for this is that the bird has beliefs and desires (or episodes of conscious thought). Therefore, the bird has beliefs and desires (or episodes of conscious thought). Griffin takes for granted that what needs to be explained is why the plover's behavior in so many different circumstances is appropriate to the end of protecting her young.

Griffin is well aware that there are explananda other than the appropriateness of the plover's behavior. He insists, for example, that there must be a perfectly good neurological explanation, which makes no reference to intentional states, for why the bird moves its wing just so at just this time. He is also well aware, with a philosophical sophistication that is rare in a scientist, that there is a perfectly good evolutionary explanation, one that makes no reference to intentional states, for why plovers, as a genus, are prone to broken wing displays. Griffin implicitly insists, however, that these explanations do not by themselves explain why the behavior of *this* plover, on this occasion, is so versatile in its adaptability. For an explanation of that, Griffin suggests, we need to appeal to *its* beliefs and desires.

In saying that the bird's behavior is versatile in its adaptability in changing circumstances, Griffin is suggesting that the bird's behavior displays a complex pattern. To say that the behavior is adaptive is to imply that the behavior varies as a function of two factors. In general, the bird's behavior will differ if circumstances in its environment differ. If the intruder approaches the bird, it will act in one way; if the intruder doesn't approach, it will act in another way. But the details of the differences in the behavior are determined by more than what is happening in the environment. The prior state and situation of the bird are relevant to what the bird does. If the bird is a nesting mother, what she does will, in general, tend to increase the distance between the intruder and the nest. The plover will approach the intruder if the intruder is not advancing toward her and away from the nest but will retreat if the intruder is following her away from the nest. If the bird is not a mother, then what it does will, in general, tend to increase the distance between the intruder and the bird itself. We need to say that the plover's behavior tends "in general" to have these results because, after all, the plover's behavior isn't infinitely adaptable. The behavior of the intruder is likely to demonstrate adaptability as well—sometimes the plover fails and gets eaten or loses her young.

For the ethologist to say that the behavior is versatile and adaptive implies more than that the bird's behavior falls into such a pattern, however complex that pattern is. The results that this versatile behavior tends to bring about—increasing the distance between the plover and the intruder in the case of the nonmother plover and increasing the distance between the intruder and the nest in the case of the mother plover—themselves display a characteristic essential to the description of the behavior as adaptive. These results are "good for" the plover or her young in a quite straightforward and uncontroversial sense. What it is to be a plover involves being an organism of a certain type. And organisms essentially are entities that, in general, are organized and behave in ways that tend to perpetuate those very ways of being organized and behaving. In this context, to say that a result of the behavior of the organism is good for some organism (the agent or some other) is simply to say that it is a result that tends to perpetuate the continued organization of that organism or, as it often is phrased, tends toward its survival. In this sense, what the nonmother plover does is good for it, and what the mother plover does is good for her young. Since versatile behavior that displays the pattern described above counts as "adaptive" if it tends to be good for the agent or, in cases of kin altruism, for the agent's kin, the plover's behavior counts as adaptive.

The adaptability of behavior is presented by the ethologist as a datum. As a matter of fact, the plover's behavior displays the complex pattern described above. The behavior, in general, does vary across a surprisingly wide range of environmental circumstances so as to increase the chances of the survival of the plover's young. Furthermore, the behavior displays novelty and versatility. Sometimes plovers demonstrate adaptive behavior in circumstances that no plover has ever faced before. The ethologist can present the versatile adaptability of the plover's behavior as a datum because that adaptiveness is a characteristic of the *behavior* itself, not the plover. The *behavior* displays the pattern of tending to change so as to bring about a result that varies as a function of the actual environment and the prior state of the plover and serves to continue the existence of the plover's young.

But now consider what an armchair ethologist, that is, a philosopher, has to say concerning whether the broken wing display, by itself, counts as appropriate behavior for protecting young birds.

> Is it appropriate for the hen not to run from the fox? This depends. It depends, among other things, on whether the hen recognizes the fox, on whether she wants to protect her chicks, on what her purposes are. Independently of these factors, the hen's behavior is neither appropriate nor inappropriate. To describe the hen, for example, as engaging in diversionary tactics (to protect her chicks) is already to describe her behavior in a way that presupposes an intentional structure for the internal source of that behavior. The appropriateness of response, then, insofar as this is relevant to what the organism believes and intends, is a property the response acquires only in virtue of its production by internal states having content.[3]

Fred Dretske, the philosopher who expounds on the significance of this analogue to the broken wing display in the above quote, more or less takes for granted that the whole notion of appropriateness of behavior, and thus the whole notion of the "versatile adaptability of behavior to changing circumstances," depends on the behavior's having been produced in the right way by internal states of the agent that have intentional content. He is interested in this assertion because he is ultimately interested in arguing that it is impossible to understand what it is for an intentional state of an agent to have the content that it does simply in terms of the role of that state in explaining the versatile adaptive behavior of the agent. Given the ethologist's description of the *behavior itself* as purposive, one might think that the intentional states of an agent that explain the behavior, the agent's beliefs and desires, borrow their contents from the purposes of that behavior and the conditions under which the behavior is appropriate for those purposes. But if *what it is* for the plover to act appropriately to protect her young is for her to act in a way that is caused by her desire to protect her young and her beliefs (1) that the young are endangered and (2) that engaging in the broken wing display will help to protect the young, then it can't be the case that what it is for the plover to want to protect her young is for the plover to be in a state that would explain her engaging in behavior appropriate for protecting her young if she were also in the belief states mentioned, where what it is to be in those belief states also depended on *their* roles in the explanation of appropriate behavior.

It is not surprising that Dretske takes for granted that what it is for behavior to be appropriate presupposes a prior concept of the content of internal intentional states. "Appropriate" is a relational qualifier having three places. A bit of behavior is appropriate in a given situation to attain a particular end. In most ordinary human contexts, when one says that some behavior is appropriate, one implies that it is appropriate in a given situation as the agent believes it to be to attain the end that the agent desires to attain. If my flipping the switch is appropriate, then it is so because I want to turn on the light and because I believe that flipping the switch will turn on the light. So it is natural to believe that if the agent had no beliefs or desires, then nothing it did could be appropriate or inappropriate, for in that case there would appear to be no grounds for fixing a goal for which the behavior was appropriate and no grounds for fixing the conditions under which the behavior counted as appropriate, as "the current state of the world" is too indefinite to supply the specific context required.

There is a second, historical, reason why it seems natural to Dretske to take for granted that behavior cannot be appropriate or inappropriate in the absence of the intentional internal states of the agent and thus that the content of the mental intentional state cannot be borrowed from the goals and presuppositions of appropriate behavior (the term that Dretske uses to describe the position of his opponents). In the middle third of the twentieth century, many scientifically minded philosophers thought that one could reduce attributions of intentional states to attributions of sets of dispositions to behave. Because these logical behaviorists were reductionist physicalists, they were vitally concerned with articulating what it is to have a mental state in terms of dispositions to behave in ways that were described in purely physical terms. As it turned out, there were systematic reasons why this proved to be impossible, for every disposition articulated solely in nonintentional descriptions fails to capture the sense of the mental term that is to be reduced. Suppose, for example, that someone suggests that for me to want to turn on the light is for me to be disposed to flip the switch if there is one in the area. But, of course, I can want to turn on the light and not flip the switch even if it is present if I don't notice it or don't understand the workings of switches. So my wanting to turn on the light can't be the disposition suggested. Logical behaviorism failed because it is inevitably circular. The behavioral dispositions that are meant to articulate what it is for an agent to have an intentional state must be specified by conditions that are themselves described in intentional terms. Given the systematic failure of logical behaviorism, it is natural for Dretske to think that *any* attempt to cash out what it is for a state of an agent to have intentional content in terms of its relation to behavior, however described, must also be circular.

By itself, nothing that Griffin presupposes when he takes the versatile adaptability of plover behavior as a datum contradicts Dretske's understanding of the relation between the content of intentional states and the appropriateness of behavior. Griffin certainly isn't suggesting that the plover's beliefs and desires are themselves adaptive behaviors: the adaptive behavior is one thing, and the beliefs and desires that Griffin thinks are necessary to explain the adaptiveness of behavior are quite another. Nor need Griffin be committed to the position that Dretske explicitly denies, that it is possible for one to specify what it is for behavior to be appropriate without making reference to the beliefs and desires of the agent, although this denial is suggested by Griffin's pattern of explanation. For, it might be maintained, the pattern of Griffin's argument requires only that there be recognizable behavioral grounds for identifying the presence of intentional states in agents, not that it be possible to specify what it is for behavior to be appropriate without reference to the intentional states of the agent of the behavior. It could be that what it is for behavior to be appropriate is for it to be caused in the right way by the corresponding beliefs and desires of the agent, although one can have good behavioral grounds for asserting that some behavior is so caused despite the lack of any independent evidence of the presence of those beliefs and desires.

Ethological data would rule out such a possibility only if there were clear cases of behavior that was, in general, adaptive and in that sense appropriate for achieving some goal but that did not warrant attributions of intentional states to the agents of that behavior. But, of course, such behavior is the rule, not the exception, in ethological description. Consider, for example, Dean Wooldridge's description of the female *Sphex* wasp made famous by Dennett.[4]

> When the time comes for egg laying, the wasp *Sphex* builds a burrow for the purpose and seeks out a cricket which she stings in such a way as to paralyze but not kill it. She drags the cricket into the burrow, lays her eggs alongside, closes the burrow, then flies away, never to return. In due course, the eggs hatch and the wasp grubs feed off the paralyzed cricket, which has not decayed, having been kept in the wasp equivalent of deep freeze. To the human mind, such an elaborately organized and seemingly purposeful routine conveys a convincing flavor of logic and thoughtfulness—until more details are examined. For example, the Wasp's routine is to bring the paralyzed cricket to the burrow, leave it on the threshold, go inside to see that all is well, emerge, and then drag the cricket in. If the cricket is moved a few inches away while the wasp is inside making her preliminary inspection, the wasp, on emerging from the burrow, will bring the cricket back to the threshold, but not inside, and then will repeat the preparatory procedure of entering the burrow to see that everything is all right.[5]

The difference between the behavior of the wasp and the behavior of the plover is in the details. Like the plover, the wasp acts so as to promote the welfare of her young. But the range of circumstances in which she can act adaptively is much more severely restricted than the range of circumstances in which the plover can respond appropriately. In general, the plover does whatever is necessary to adapt to changing circumstances so as to promote the welfare of her young, although there are, of course, limits to this adaptability. The behavior of the plover seems more rational than that of the wasp and thus gives us more grounds to think that the plover has intentional states and acts for reasons, just to the extent that the behavior of the plover is more flexibly adaptive in the face of a wider range of changing environmental circumstances than is the behavior of the wasp.

No one thinks that the wasp thinks, that it has intentional states that have content. But it is quite plausible to think that the plover has intentional states that have content. But the difference between the behaviors of these two organisms is not that what the plover does is adaptive in a versatile way in changing circumstances and what the wasp does is not adaptive in a versatile way in changing circumstances. If what the wasp did in changing circumstances were not adaptive in this way, there would not be so many wasps. What the plover does is *more* versatile and adaptive than what the wasp does in a wider range of circumstances. The plover can deal with the presence of experimental intervention far more effectively than the wasp can, for example. And the plover is far more capable of adjusting her behavior in light of what seems to be a recognition of the failure of a previous behavior to achieve its proximate end than is the wasp. You won't find a plover endlessly repeating a failed subroutine in the way that the wasp in the example does. For those reasons, among others, we say that the plover's behavior is more *rational* than that of the wasp and we have better, or at least some, reason to think that the plover has intentional states. But for all that, what the wasp does in a wide range of circumstances is appropriate in the sense of being good for the wasp and its kin. Wasp behavior displays teleology, if not rationality.

The fact that wasp behavior, for example, displays this adaptive pattern even in the absence of rationality provides the rudiments of a response to Dretske's concerns. Dretske assumes that no behavior can be appropriate or inappropriate apart from the intentional states of the agent of the behavior. But the very notion that behavior itself can be adaptive or maladaptive presupposes that there is a sense of "appropriate" other than the one that is common in intentional contexts. In this other sense, "appropriate" remains a three-place relation in which one place is occupied by the behavior, the second place is occupied by what the behavior is for or the goal of the behavior, and the third place is occupied by some variant on a specification of current conditions. But in this sense of "appropriate," what determines the goal of the behavior—what the behavior is for, the end in relation to which it can be appropriate or not—is not what the organism wants, but what the organism *is.*

Rational behavior differs from behavior that is merely adaptive and has a goal insofar as it is more flexible and better capable of coping with novel situations. Both of these qualifiers, "more flexible" and "better able to cope with novelty," presuppose that the behavior has a goal. In the case of rational behavior, however, the variety and flexibility of the behavior itself requires that this goal be described in more detail than in the case of nonrational adaptive behavior. To know what a mother *Sphex* wasp will do on a given occasion, it is sufficient to know that it is a *Sphex* wasp and what it is that *Sphex* wasps do in that situation that is adaptive for facilitating the development of their young. But insofar as the behavior of the plover displays rationality, this sort of knowledge is insufficient to explain her behavior. In addition, one needs to know what it is that the plover is trying to achieve right at the moment, even if one can safely assume that, whatever the plover's current goal, it serves the ultimate end of helping her young. (Remember, the plover's behavior is adaptive.) We mark this difference by the familiar sense of "appropriate," in which behavior is appropriate or inappropriate in relation to an end that is specified by a factor specific to the individual agent, what it is that the agent desires.

Intentional states, such as beliefs and desires, are states that are about or directed toward something. At least some such states, again including beliefs and desires, are states of agents that help in giving rational explanations of what the agent does. When I flip a light switch, I may do so *because* I want the light to be on and because I believe that my flipping the switch will bring this about, so my belief and desire explain my flipping: if I act on this desire and this belief, then they constitute my reason for acting. As the invocation of the notion of a reason in this context makes evident, not all behavior admits of explanation in terms of intentional states. In general, we give rational explanations by appealing to the intentional states of an agent, such as its beliefs and desires, only if that behavior itself is rational.

It is pretty clear that this explanatory relation between intentional states and rational behavior is crucial to our ordinary ways of attributing intentional states. Aside from first-person introspection, the only way we can judge whether some entity has intentional states is by observing that entity's behavior and noting that it is rational. But when does the behavior of some agent count as rational and thus constitute evidence that the agent is acting on reasons and, therefore, evidence that the agent has intentional states such as beliefs and desires? It would seem that the behavior of an agent must display, at a minimum, "versatile adaptability to changing circumstances and challenges" if that behavior is to count as rational. When one acts in an instrumentally rational fashion, one does those things, and only those things, that one has reason to believe tend to bring about one's ends.[6] But if those ends are not, in general, conducive to the welfare of the agent or if what the agent has reason to believe is true is not generally true, then the agent's behavior does not count as rational. The fact that rational agents generally act in their own interests guarantees that rational behavior is adaptive, and the fact that what rational agents believe is generally true guarantees that rational behavior is versatilely successful in the face of changing circumstances and challenges. So to count as rational, behavior must be versatilely adaptive to changing circumstances and challenges. Rational behavior is a species of adaptive behavior, and we, as rational agents, are rational animals.

The fact that Dretske is wrong in thinking that behavior can be appropriate only when it is explained by the agent's intentional states—that is, when it is rational—suggests that he is also wrong in his overall claim. Dretske and many others think that it is impossible for the intentional content of mental states to be borrowed from the goals that the behavior that these mental states help to explain is appropriate for attaining. The usual reason cited for this conviction is the one mentioned by Dretske: to appeal to the goals of appropriate behavior in an articulation of what it is for a mental state to have an intentional content is circular because behavior can have a goal for which it is appropriate only if it is caused by mental states with content. But ethology reminds us that there is a perfectly acceptable nonmetaphorical sense in which behavior *can* be appropriate for attaining a goal even in the absence of mental states, the sense in which the wasp's behavior is adaptive. So it is at least possible that the contents of intentional mental states derive from their roles in the rational explanation of goal-directed behavior.

In this book I develop an account of intentional content that works out this possibility. This account, which I call a pragmatic account, asserts that the contents of mental states derive from their roles in providing a certain kind of explanation for behavior described in a certain way. Just as Griffin's argument assumes, what intentional states explain is why a particular agent acts in a particular, adaptive, goal-directed fashion. That is, the explanandum of any explanation that invokes beliefs and desires is behavior that is described as being adaptive toward reaching a goal. The explanations of these goal-directed behaviors also display a distinctive structure. They are rational explanations in which the states of the agent that are invoked, and their relations, stand under the constraints of rationality: consistency, coherence, and charity.

Any such pragmatic account of the contents of intentional mental states must meet two requirements. First, since a pragmatic account of intentional content understands the content of mental states by appealing to the roles of those states in explaining goal-directed behavior, the account must specify, independent of any reference to the mental states of the agent, what it is for behavior to have a goal. This task is accomplished in the first four chapters. Second, because the account suggests that rational behavior is a species of goal-directed behavior for which it is appropriate to offer rational explanations, one must give a general account of rational explanation and articulate how the contents of mental states depend on their roles in such rational explanations. This is the task of the remainder of the book.

CHAPTER ONE

# WHAT IS INTENTIONALITY?

## 1. THREE CENTRAL ISSUES CONCERNING INTENTIONALITY

This is a book concerning intentionality. The question that it attempts to answer is straightforward, if difficult: what sort of feature must an event, state, or entity have if it is to be about something or directed toward something, that is, if it is to count as intentional? In this chapter, I prepare the ground for my answer to this question by reflecting on the character of the question itself. In the course of this reflection, we will discover three issues that any account of intentionality must address. The adequacy of the responses to these issues in any account of intentionality can thus serve as a criterion for judging the adequacy of that account.

The question of intentionality is not a question concerning minds. Minds are the wrong sort of entity for the inquiry I intend. *Minds* aren't about or directed toward anything, even if some, or all, states that are about or directed toward something are states of minds. When I think about my son or believe that my daughter is in Massachusetts or walk to the store, it is not my mind that is intentional. It is my thought that is about my son, my belief that concerns my daughter, and my walking that is directed toward the store. In general, it is states or events that exhibit intentionality, and minds are neither states nor events. They are persisting entities.

The question of what feature an event or state must display in order to exhibit intentionality is, in some respects, quite odd. The answer would seem to be trivial. In my initial statement of the issue, I specified what it is for some state or event to be intentional. Some event or state is intentional just in case it is about or directed toward something. This specification amounts to a definition of intentionality, an answer to the question of what it is for a state or event to count as intentional. So the question I mean to raise can't be the question of definition.

Of course, there are other questions that one might raise about intentionality that are not answered by this definition. For example, each state or event that counts as intentional presumably has some cause or other, and the whole class of such states may have been caused to come into existence at some point. So there are several causal questions in the area. But none of these appears to be a promising topic for philosophical inquiry.

I do not intend to consider the question of the grounds on which we attribute intentionality. We are pretty good at recognizing examples of intentional states, so we have a pretty good implicit working knowledge of the criteria to be used in making that attribution. Furthermore, it is obvious when we are attributing intentionality because we have a good linguistic test for the presence of such an attribution, a lack of substitutivity salva veritate. While there are interesting issues involved in making explicit the grounds on which we are warranted in asserting that some state or event has intentional content—which we will consider later—this task is not my focus.

Instead, the focus of this inquiry is ontological. While we have an intuitive sense of what it is for an event or state to be about something or directed toward something and some workable criteria for attributing these properties, we do not have a clear apprehension of what these characteristics of events and states *consist in.* This lack of understanding is indicated by the existence of at least three separate respects in which we find these characteristics, aboutness and directedness, mysterious. Any account of what it is for a state or event to be intentional, that is, about or directed toward something, must dispel each of these mysteries.

Of these mysterious aspects of aboutness, only one is peculiar to our time. That issue concerns the physicality of intentional states. It is hard to see how any merely physical thing, event, or state—that is, any thing, event, or state that admits of being adequately described in the language appropriate to physics—could also have the property of being about something. The development of science over the past 350 years has led most of us to commit ourselves to the view that every individual that exists admits of being adequately described in physical terms. So part of the mystery of intentionality concerns the presence of aboutness and directionality in a physical world, that is, a world that can be adequately described in the terms of a physics that does not mention intentionality.

But there is something surprising and wonderful about intentionality even for one who isn't a physicalist. On the face of it, for any one thing to be about some other thing involves a relation between those two things. In all usual cases of relational properties, for a relation to obtain, both of the relata must exist. If Dick is taller than Jane, both Dick and Jane must actually exist. This is not true of intentionality, and that it is not true is central to intentional relatedness. When Dick thinks about Jane, it is possible that there is, in fact, no Jane. And I can be hunting snarks when there are no snarks. So there is a second mystery surrounding intentionality, one that has nothing to do with physicalism. This mystery, which I will call the "puzzle of intentionality," is simply this: How must we understand what it is for one thing to be about another when that one thing can be about another even when that second thing does not exist at all? Is intentionality a relational property? If so, how can there be a relation that does not imply the existence of all of the relata? Is intentionality an intrinsic property? If so, how can there be an intrinsic property that implicates another entity in some way?

This logical peculiarity of intentional language is related to the third mystery regarding intentionality. A state is intentional just in case it is of, about, or directed toward something. As such, any intentional state can be thought of as involving a representation, a characterization of how things are or might be. Even acts contain within themselves a goal, or a way the world is to be as a result of the act. This representational characterization of the world is the representational content of the intentional state. But the world need not be or ever come to be as it is represented to be in an intentional state, and this possible lack of fit is essential to our notion of what it is for a state to be about something. The content of any intentional state has built in to it that it is a re-presentation, a veridical giving of the world or a giving of how the world is to become. But it is also built in to this intention as an intention that it is possible for it to fail to agree with its object, present or future. Indeed, to say that it is possible for there to be lack of agreement between the world and how it is represented is just a more general way of saying that it is possible for an intentional state to be about something that does not exist. ("More general" because a state can be about something that does exist but can represent it in a way that is not true to this thing.) Since the content of an intention can agree or fail to agree with its object and since, as intention, it *ought* to agree, such contents are subject to assessments of correctness and incorrectness. That is, intentions as intentions stand under norms; they can be evaluated as being true to their objects or failing to be true to their objects. But this evaluability is mysterious.

What is mysterious about the normativity of intentions is that their evaluability according to norms seems to be part of what it is to be an intention with intentional content. In many cases of normative evaluation, both whether something is evaluable and the norm in terms of which it is evaluated depend on something extrinsic to the item evaluated. As Heraclitus pointed out, salt water is both good and bad: good for fish and bad for us. In itself, apart from fish or us, however, it is neither good nor bad. Intentions are different from salt water. Anything, of course, can fail to be a representation, can fail to represent. Such a state, however, is not intentional. Intentions aren't like that. They *are* representations. They do represent something. It's just that they *might* not be true to what they represent. Insofar as they are intentional, then, they are evaluable; evaluability is part of what it is for a state to be intentional at all. In addition, the norm to be used in evaluating the state is implicated in the state itself. If I believe that the cat is brown, then it is the actual color of the cat that is the norm against which this belief is to be judged. And that it is the color of the cat that determines the truth status of the belief is fixed by the content of the belief itself. If an act has the goal of my being in Chicago, then my position relative to Chicago after the act is the norm to be used in evaluating the act. That it is my position relative to Chicago that determines the success or failure of my action is fixed by the goal of my act itself. An intentional state carries within itself both the fact that it is to be normatively evaluated and the norm according to which this evaluation is to be made. How this is possible is the third mystery regarding intentionality.

When I ask what it is for an event or state to be about or directed toward something, and thus to be intentional, I am asking for an ontological account of aboutness. As I am using the word, an account of intentionality must address the three mysteries. The first mystery concerns the relation between intentions and the nonintentional: how does intentionality fit into a physical world? The second concerns intentionality itself:

how, if at all, is an entity related to something it is about or directed toward? The third mystery concerns the normative status of intentions: how is it possible for intentional states to be inherently evaluable regarding their correctness or incorrectness?

To address these mysteries, an account must go some way toward dispelling them. It must first give us an understanding of what is involved in intentionality that displays how it is possible for things that are adequately describable in physical terms also to be describable in intentional terms. Such an understanding might involve a reduction of the intentional to the physical, either through an analysis of the intentional vocabulary in physical terms or through the kind of "theoretical definition" (to use Ruth Millikan's phrase) involved in the identification of heat and mean kinetic energy. But it need not. It might, for example, turn out that the intentional realm is autonomous and neither analytically nor theoretically reducible to the physical but that, given the understanding of intentional properties given in the account, there is no particular problem with things that are physical having them. This is essentially the character of the various functionalist accounts, for example.

An account of intentionality must also go some way toward dispelling the mystery concerning the logical status of intentional predications. The basic features of that status are plain enough. A can be about B even when B doesn't exist. Substitution of coreferring expressions in intentional contexts does not always preserve truth-value. The mystery concerns the sort of properties that could have these features. What marks intentional properties as unique is that most other properties, either intrinsic or relational, entirely lack these features. An account of intentionality should make intelligible why the structure of intentional language is so odd.

Finally and most importantly, an account of intentionality should make intelligible why and how intentional states stand under the norms that govern them.

These three requirements of an account of intentionality interact with one another. The understanding of the place of intentionality in nature should make the logic of intentional language understandable and the normative assessability of intentions intelligible. At a minimum, the account of how intentionality relates to physicality should not have implications that flatly contradict the ordinary logic of intentional discourse. I will take the adequacy of the treatment of these three mysteries by an account of intentionality as the prime criterion for the adequacy of that account. Taken as a whole, this book presents the argument that a particular approach to intentionality, which I will call the pragmatic approach, does a better job of handling these issues than any of its more popular competitors. On those grounds, I will recommend this approach to the question of intentionality.

## 2. THE PRAGMATIC APPROACH TO INTENTIONALITY

There are two kinds of entities to which we are tempted to attribute intentional content. On the one hand, we say of persisting entities, persons and perhaps certain animals, that they have beliefs, desires, and so on, and it is pretty clear that where there are such attitudes there is also intentionality. Persons and animals are entities that persist through time, and their beliefs and so on are states of these entities that might also persist over time. On the other hand, some *acts* of such persisting entities, most obviously linguistic utterances, clearly have an intentional character insofar as they are meaningful. But much human nonverbal behavior shares with the behavior of other animals the feature of being appropriate, in some circumstances, to the achievement of some end, and therefore seems to be directed toward the achievement of that end. In these linguistic and nonlinguistic cases, it is the behavior itself that has the property of being about or directed toward something. When my two-year-old son said, "That's milk!" believing that "milk" referred to Coke, *he* believed that what he was saying was that the thing over there was Coke. But the sentence he uttered meant that there was milk over there.

We characteristically explain the occurrence of one of these kinds of states, the acts, by appealing to the occurrence of the other, the beliefs, and so on. So it seems natural to explain the intentionality of one kind of state by appealing to the intentionality of the other.

It should be clear from the beginning that there are two quite distinct questions regarding the relation between thoughts and actions. First, what is the explanatory relation between distinct occurrences of thoughts and actions? I take it to be obvious that in very many standard cases the thoughts of an agent directly explain its acts but that, at best, how the agent acts only indirectly influences its thoughts. The fact that I want the light to be on and that I believe flipping the switch will result in the light's being on directly explains my flipping the switch. But my flipping the switch doesn't directly explain any of my beliefs and desires.

There is, however, a second question concerning thought and action. What is it for a thought or an action to be about or directed toward something and, in virtue of that directedness, to have the intentional content it does? The explanatory connection between thought and action makes it natural to think that the correct way to answer this second question is by dividing it. This strategy directly answers the question of the nature of content for only one element of the action/thought pair and then explains the fact that the other element of the pair also has content by in some way alluding to the relation between them. This general strategy, and the fact that the occurrences of thoughts explain the occurrences of actions, leaves open, however, the question of which element of the thought/action pair should be seen as having original content and which as deriving its content (i.e., whether it is the intentionality of thought or the intentionality of action that is explained by reference to the other).

To some philosophers, the fact that the occurrence of an agent's thoughts explains the occurrence of an agent's acts suggests that the intentionality of those acts should also be explained by the intentionality of those thoughts, rather than the other way around. This intuition leads to the attempt to specify what it is for a mental state to have intentional content independent of the role of such states in the explanation of goal-directed behavior. Once this is done, it seems an easy task to account for the intentionality of action as derivative from the intentionality of the internal states that cause it. Roughly, in this view, some act is performed in order to achieve some end just in case it is partially caused (in the right way) by a desire to achieve that end.

The fact that the intentional states of agents explain the occurrence of the goal-directed behavior of those agents strongly suggests that this is the right strategy. It is certainly the strategy followed by most twentieth-century attempts to give an account of intentionality. But the history of these attempts has shown that it is devilishly difficult for either of the two main programs for understanding the intentionality of thought to give an account of the semantics of thought. The problem is that whether one appeals to a naturalized form of representation relation to fix the intentional content of thoughts or to the causal roles of mental states, one must account for the normativity of intentionality, the fact that representations must be capable of misrepresenting and beliefs must be capable of falsity. Neither of the standard strategies for understanding intentionality has been notably successful in accounting for this possibility. This problem suggests to some of us that the intentional content of thoughts should be understood as deriving from their relation with actions, rather than the other way round.

The obvious problem with treating the goal-directedness of action as the basic form of directedness, rather than treating the aboutness of thought as basic, is that in most human contexts we attribute a goal to an act only if it is intentional, that is, caused in the right way by the intentions, beliefs, and desires of agents. But we are quite willing to attribute to human agents thoughts that do not cause behavior. This suggests that behavior has a goal only if it is caused in the right way by intentional states of agents. If this is the case, it is circular to account for the intentionality of thought by appealing to its relation with the directionality of behavior.

This appearance of circularity has begun to dissipate in the last forty years, however, largely as a result of developments in the philosophy of biology. A succession of philosophers, most notably Charles Taylor, Larry Wright, and Ruth Millikan, have gone a long way toward making teleological language intelligible in a way that neither reduces it to physical terms nor depends on any causal relation between the items displaying the teleological features and the intentional states of agents.[1] If behavior can have a goal independent of its causal relations with the intentional states of agents, as this work suggests, then there is no prima facie reason to think that it is impossible to derive the aboutness of mental states from their explanatory relations with goal-directed behavior.

There are two main kinds of teleological properties. We speak of both continuing entities and behavior as having functions, and we speak of behavior as having goals. Taylor and Wright offered helpful, if not entirely adequate, accounts of both types of teleological properties. To this point, however, only their discussions of functions have been applied to questions concerning intentionality, most notably by Millikan. Millikan's overall strategy is to understand the representational content of beliefs and desires in terms of the biological functions of the states that instantiate them. Her view thus derives the aboutness of mental states from the functionality of types of biological entity. But no one as yet has attempted to exploit the independent goal-directedness of behavior, which is also suggested by Taylor's and Wright's work, to develop an account of aboutness in terms of the directionality of behavior. This is the overall strategy of this book.

According to such a pragmatic account, the intentionality of thoughts is to be understood by way of the relationship of those thoughts with the actions that they could rationally explain. As the example of the *Sphex* wasp displays, not all goal-directed action is rational, and it is only rational action that needs to be explained by an appeal to the agent's reasons. Behavior that has a goal is rational only if it responds appropriately to a wide range of varying and novel conditions so as to achieve its goal. Such behavior cannot be explained simply by appealing to what it is that all organisms of a certain type do in a given situation so as to achieve their ends. The actions of the wasp fail to meet this standard. The individual wasp's behavior is fully explicable once one understands that it is a wasp of the type that it is, for all such wasps act in the same ways in the same situations to achieve the same goals. We do not need to attribute idiosyncratic, particular goals to such individuals in order to understand why their behavior has the goals that it has. That is, we have no grounds for attributing desires to such creatures and, hence, no grounds for attributing beliefs either. Only when an individual develops a novel strategy for coping with a novel situation do we see the behavior as rational and attribute reasons to that individual. Those reasons are the explanans of rational explanations of the agent's rational, goal-directed, behavior. These reasons are particular to the individual agent—that is, they are states of that agent.

In the pragmatic view presented here, these are the intentional states of the agent. Which state such a state is, is fixed by the teleological character of the acts that it helps to explain and the particular role that it plays in these rational explanations. Since intentional states are typed and individuated by their contents, in the pragmatic view the contents of intentional states are fixed by the teleological character of the acts that they help to rationally explain and their roles in those explanations. If what the plover does is rational, then the plover wants the intruder to follow her. That is, roughly, the plover is in a state such that if she wants to protect her young and believes that if the intruder follows her, this will help protect the young, then it *should* also be in this state; and if the plover is in this state and believes that getting the intruder to believe that she is injured will cause the intruder to follow her, the plover *should* do something to induce the intruder to believe that she is injured. Needless to say, all of the intentional descriptions used in this partial definition need to be cashed out in nonintentional (but perhaps teleological), terms. The normative modal term "should" is required because intentional states are factors in *rational* explanations, and rationality requires the possibility of error. As we will see, the use of such language also implies that rational creatures, that is, creatures with mental intentions, mostly act, believe, and desire as they should.

Such a position, which emphasizes the importance of action for the understanding of intentional content, has at least two noteworthy virtues. First, and most obviously, the only evidence that we have for the presence of mental states in our fellow creatures is what they do, when what they do has some point. Even if one isn't a verificationist who simply identifies what it means to say that "S believes that P" with the evidence in favor of asserting that "S believes that P," it would still be a virtue if the contents of mental states were public in the sense that they could be detected by an observer situated as each of us is in respect to others.

There is a second virtue to the pragmatic view of content that is perhaps a bit less obvious. Such an account makes intelligible the relation between intentionality and rationality. Our mental states are thought to be about something and to be rational, in the sense of being related to each other and to our actions in such a way as to provide reasons. If one has the kinds of mental states that we think ourselves to possess, then there are reasons why one does what one does and thinks as one thinks. For some state of an individual to count as a reason, it must be related to a set of other possible states of that individual in such a way that if the entity has the first state (and perhaps some others), then the entity *should* have some of the others. If I believe that Gray is between Portland and Lewiston and that Portland and Lewiston are thirty-five miles apart, then I should also believe that Gray is less than thirty-five miles from Portland. Similarly, if I want to go to Gray and believe that Route 26 is the best way to get there, then, all other things equal, I should take Route 26. But why should it be true that the same states that are about something also count as reasons? The pragmatic view of content provides a ready answer to this question. Mental states have content *in virtue of* their possible role in rationally explaining actions, so it follows that a state has content *only if* it could provide a reason for performing an action.[2]

So there are some philosophical advantages to developing a plausible action-oriented account of intentional content. Nevertheless, this approach clearly has not dominated discussions of intentionality. Instead, the two programs that have dominated the landscape for the last forty years have developed the alternative strategy of accounting for the intentionality of action by appealing to the intentionality of thought, although they have done so in different ways. Indeed, it is fair to say that these views have been so dominant that most philosophers simply have not found any reason to explore other alternatives (that is, beyond eliminativist junking of the very notion of intentionality in the name of explanatory efficiency). To better understand the pragmatic approach I champion, it will be helpful to briefly contrast it with its more familiar competitors.

## 3. REPRESENTATIONALISM, FUNCTIONALISM, AND PRAGMATISM

The two dominant programs for understanding intentionality agree that the intentionality of thought is to be used in explaining the intentional character of action. One of these strategies, the representationalist alternative, begins with the recognition that every mental intentional state involves a specification of a possible way the world might be. When we say that Jane believes that the sky is blue or hopes to be home in time for dinner or wants to eat fish for lunch, our attributions of those states include a clause that appears to articulate a possible state of the world as well as a verb that specifies a relation between Jane and that possibility. So it is natural to think that the states themselves that we refer to in these attributions also involve two components: a representation of the world and an attitude of a subject toward that representation. But if A represents B, then A is about B, regardless of whether some subject stands in some relation to A. So representationalists such as Jerry Fodor, Ruth Millikan, and Fred Dretske attempt to account for the intentionality of thought by giving a naturalized account of the representation relation.

Representationalism has another root, a root that goes back at least as far as Descartes. Descartes fixated on the second mystery concerning intentionality. The Cartesian problem in epistemology arises through its focus on the fact that it is always possible to intend an entity that does not exist or to intend an entity that does exist but to intend it as being in a way in which it isn't. But real relations imply the existence of all of the relata. This implication has led numerous philosophers from Descartes to Fodor to introduce a third entity to mediate between the subject and the possible state of affairs toward which it has an attitude. However much ideas differ from instantiations of propositions in mentalese, each tertium quid shares two features that derive from the problem it is meant to solve. First, these third entities actually exist. There is no point in positing an entity to mediate a relation between a subject and a possibility if that entity is itself only possible. Second, they have a distinctive relation to the possible state of affairs toward which the subject has an attitude: they represent it. There is no point in positing an entity with which a subject can be related unless that entity itself stands in relation to the possibility toward which the subject has an attitude. What is distinctive about the representational relation, however, is that if A represents B, B need not exist. Only in this way can a representation mediate between a subject and a possibility.

Representations are real entities that stand in two sorts of relations: they represent possible states of affairs, and they are the objects of propositional attitudes. What makes them representations, and what accounts for the intentionality of any intentional state, is that these items represent possible states of affairs. For the representationalist, it might very well be the case that for some subject to have some attitude toward some representation essentially involves the role that that representation plays in the rational explanation of action but the *intentional* character of the belief, or whatever, is a function solely of the representational character of the item toward which the attitude exists.

Functionalism, on the other hand, has a different orientation. For the functionalist, the fact that mental states explain behavior provides an anchor that can be used to fix the content of those states. As we ordinarily explain the fact that Jane opens her umbrella by appealing to Jane's beliefs concerning umbrellas and staying dry in the rain, together with her desire to stay dry, and we explain the fact that Jane believes that it is raining by alluding to what she sees out her window, the functionalist defines what it is to have those beliefs and desires in terms of their roles in the possible explanation of behavior and in terms of the sorts of sensations that are most directly linked to specific beliefs. For the functionalist, "each type of mental state is a state consisting of a disposition to act in certain ways and to have certain mental states, given certain sensory inputs and certain mental states."[3] That is, for a given set of mental states and sensory inputs, each mental state is identified with a causal disposition to act in a certain way and to have other mental states. So, whereas a behaviorist might identify the belief that it is raining with the disposition to open an umbrella, the functionalist might partially specify the same belief as the state B such that (1) if an agent were in state B and believed that opening an umbrella would keep her dry when it is raining and further desired to stay dry, she would be disposed to open the umbrella; and (2) if the agent were in sensory state S and believed she were awake and outside, she would be disposed to be in state B. Functionalists see mental states as intentional because they belong to a certain sort of system for explaining behavior on the basis of inputs. And, insofar as that behavior itself is seen to have an intentional character, it is only in virtue of its being explained by that kind of system.

One way to understand the relation between functionalism and representationalism is historical. Seen from that perspective, representationalism is not so much a competitor of functionalism as an adjunct to it. Causal-role functionalism began as an explanation of how a merely physical entity could also be a mind. The proposed answer took its lead from issues of rationality. How could a merely physical system behave in ways that mimicked or displayed rational thought? The functionalists proposed that a system organized so as to embody a set of causal dispositions could, in principle, be constructed so as to qualify as rational. The only real problem was to determine the particular dispositions, or programs, themselves. Insofar as functionalists identified mental states with causal dispositions, however, they ran into intractable problems concerning the intentionality or semantics of those states. Briefly stated, the problem was that if one uses the causes and effects of a state to fix its type and if that type is identified with the content of, say, a belief, then it becomes hard to see how the belief could ever be false. The logic of rational explanation demands that the effects of a belief on behavior and on other mental states are the same whether that belief is true or false, so the effects are not available to distinguish true and false beliefs. So the functionalist is thrown back on causes. But any time a token of the type that is identified with the belief occurs, it must have *some* cause, and that cause is necessarily an instance of one among the types of events that cause tokens of that type. So the content of the belief is a disjunction among all of the types of events that cause tokens of the belief type. This is the dread disjunction problem.

The crux of this problem is the normative dimension of intentionality. It is necessary to beliefs as beliefs that they can be evaluated regarding their truth or falsity and that it be possible that at least some beliefs are false. But if what it is for some type of belief to have the content it does is fixed by the disjunction of all the kinds of events that can cause mental states of that type, then it is hard to see how it ever could be possible for any belief state to be false. For any instance of such a belief state must have some cause or other, and ex hypothesi the cause of that state would be part of the disjunction that fixes the content of states of this type.

One way to respond to this problem is to divide the issue of rationality from that of intentionality. For a system to be rational can still be understood as operation according to a program. (We will see that there are problems even for this characterization of rationality.) The semantics of mental states, however, is seen to be a product of something other than the causal role of those states. In particular, the intentional content of a state type is identical with what that type represents. This is the birth of representationalism. The only problem is the difficulty of articulating what it is for something to represent a possibility.

The pragmatic approach to intentionality that I defend differs fundamentally from the representational strategy. For a pragmatist, the primary intentional events are goal-directed actions. Mental states have their intentional content conferred on them through their role in the rational explanation of those acts. For the representationalist, on the other hand, the intentional character of mental states is, in principle, separable from their explanatory role. For representationalists, *content* is first and foremost a matter of representation, and in order to understand intentionality, one must understand how one entity can represent another.

The pragmatic approach shares with functionalism an emphasis on the explanatory role of mental events for an understanding of intentionality. It differs from functionalism in three related and fundamental respects, however. First, orthodox functionalism "insists that characterizations of mental states should contain descriptions of inputs and outputs in physical language,"[4] whereas the pragmatic approach insists that it is important that the behavior that is the explanandum should admit of a goal-directed description independent of its being caused (in the right way) by mental intentional states. As an outgrowth of behaviorism, functionalism developed as an attempt to give a nonintentional specification of the propositional attitudes. To do this, it is necessary that the dispositions that form the core of that account should not be infected with intentionalistic features. The pragmatic approach, on the other hand, not only has no interest in physicalistic reductions but also sees the intentional character of mental states as derivative from their role in explaining goal-directed action, which in turn is given a naturalistic, though nonreductive, articulation. This points to a second difference between functionalist and pragmatic accounts. Functionalism agrees with representationalism in seeing the intentionality of mental states as original and the intentionality of action as derivative. The disagreement with representationalism is merely whether the intentionality of those mental states depends on their belonging to a system of a certain sort for explaining behavior or on their including representations. Because the pragmatic approach appeals to the teleology of events to anchor its attributions of content to the states that explain them, it reverses the procedure and interprets the intentionality of thought as dependent on its role in explaining goal-directed action. Finally, perhaps because of its historical links with cognitive science and the artificial intelligence community, orthodox functionalism demands that the links among the types of inputs, mental states, and behaviors that are related in the system that is implicated in rational explanations of behavior are all causal. As we will see, the pragmatic approach suggests that at least some of these links cannot be causal because they embody normative regularities.

We can thus develop the following two-dimensional matrix for typing responses to the problem of intentionality. Either the teleology of action is explained by appealing to its relation to the intentionality of thought or the intentionality of thought is explained by appealing to the teleology of action. In this dimension, functionalism and representationalism agree in giving pride of place to thought over action, while the pragmatic approach that I develop reverses this priority. Second, either the fact that particular occurrences of our propositional attitudes explain our particular acts is essential to the proper understanding of the intentionality of thought or it isn't. In this dimension, the pragmatic approach agrees with functionalism in holding that it is the explanatory role of thoughts that is central to an understanding of their intentional status, whereas representationalism does not accept this proposition.

The key distinguishing features of the pragmatic position that I will develop are the suggestions that actions have their goals originally and that the mental states of rational agents have their contents only in virtue of the roles they play in the rational explanation of a species of goal-directed behavior, rational action. To make this position plausible, it is thus necessary to perform three tasks that are not required of either a representationalist or a functionalist theory of content. First, one must provide a plausible account of what it is for some behavior to have a goal, an account that does not appeal to that behavior's being caused in the right way by the intentional states of an agent. Second, one must provide a plausible set of differentia for distinguishing rational from nonrational goal-directed behavior. Finally, one must articulate a description of the nature of rational explanation that displays just how the contents of mental states are fixed by their roles in rational explanation of action. In the next chapter I turn to the first of these tasks, while the second and third issues are treated in chapters 5 and 6.

CHAPTER TWO

# GOALS WITHOUT INTENTIONS

## 1. THE FIELD TELEOLOGIST'S PROBLEM

Consider the plight of the field teleologist. Word has just come back from the Sirius probe that there is nothing that chemically resembles the carbon-based life with which we are familiar on the single planet orbiting the star. Nevertheless, some members of the exploration party report that it seems as though a great deal of what is happening around them falls into a pattern, a pattern in which some events happen in order to bring about other events, or have some goal. These members of the exploration team can't state exactly what the goals are, but they still think that there are *some* ends to what they see. Other members of the team disagree. After all, they have discovered nothing that in principle can't be explained without the assumption of such teleological processes. The field teleologist has been called in to resolve the resulting dispute.

The situation of the field teleologist is analogous in a variety of ways to those of the field linguist and the radical interpreter made famous by W. V. Quine and Donald Davidson respectively.[1] First, all three investigators face the task of attributing a property to behavior. Behavior, of course, is *an event* or series of events. Quine's linguist must specify the meaning or, more properly, the appropriate translation of an utterance; Davidson's interpreter must specify the truth conditions on an utterance; and the field teleologist is called on to decide the attribution of a goal to an occurrence. Second, in all three cases it is difficult to see how the property in question can be understood in purely physical terms. Goal-directedness, like appropriate translation and truth conditions, seems to imply a semantic or intentional character that is mysterious, at best, when viewed from a physical perspective. Third, the task actually performed by the scientist in each case is one that ordinary persons appear to perform effortlessly. If anything, in ordinary contexts our goal-detecting machinery is even more finely tuned than our ability to decipher what our fellow human beings are saying. And it is a good thing that it is. After all, if we could not differentiate among a nonagent, an agent that is attempting to eat us, and an agent that is trying to avoid us, we wouldn't last long.

Fourth, and perhaps most important, the conditions of the three thought experiments are designed to set up an epistemological puzzle that has, if properly understood, ontological implications. Assume that an observer comes into the world without the usual background information available to ordinary human adults. What features of events could such an observer use to determine whether some events had a meaning or truth conditions or a goal—and, if they did, to determine what that meaning, those truth conditions, or that goal was? It is quite unlikely that any actual human infant is in the position of the radical interpreter, and it is even less likely that there has ever been an interpreter who needed to rely on behavioral evidence alone to make judgments of goal-directedness. These judgments are too important to the survival of the organism for evolution not to have hardwired some of the cognitive skills that help us to make these attributions. Nevertheless, there is an ontological point to setting up the problems in this way. Because the artificial situation takes away ordinary knowledge about the events to which we attribute these odd properties—except those features that are sensible in a narrow definition—it raises the issue of what else, if anything, we would need to know in order to make the attributions in question. That is, we should focus not on *how* we could know the meaning, truth conditions, or goal of an event in these peculiar circumstances but rather on *what* we must know to be able to attribute them. So, for example, if it should turn out that in the linguistic arena the behavioral evidence is insufficient to generate knowledge of syntax and that knowledge of syntax is necessary for semantic ascription, then the thought experiment of the field linguist should help us to discover this fact. That is, the fact that the field linguist could not solve her problem unless she could assume knowledge of the syntax of her interlocutor would show us both that syntax is necessary for linguistic meaning and that syntax necessarily involves something in addition to observable behavior. Similarly, if the field teleologist could not solve her problem with the limited tools available to her, this failure would indicate that, whatever it is for an event to have a goal, goal-directedness is undetectable solely on the basis of an observation of what occurs.

Although there are important similarities between the situation of the field teleologist and those of her semantic cousins, there is nevertheless a striking difference between their problems. The problem confronting the teleologist is far more radical. In fact, both the field linguist and the radical interpreter presuppose that the field teleologist has already completed her work, but their own work is mostly irrelevant to her task. Quine assumes that his investigator can recognize assent and dissent. And, if she is to make any progress at all with her job, she must be able to do this. But both assent and dissent are *acts* of agents who have the proximate goal of communicating. So it is assumed that the field linguist has the ability to recognize at least some agents and to recognize the goals of some acts of those agents. Similarly, if the radical interpreter is to get started with her task, she must be able to recognize the attitude of holding true directed toward sentences, which is the evidence that is directly relevant to the construction of a T theory for the language of a speaker. But, as Davidson emphasizes, to do this presupposes being able to recognize a teleological pattern in the behavior of an agent.[2] So radical interpretation, too, presupposes that the job of the field teleologist has already been carried out.

There is another related respect in which the field linguist and the radical interpreter presuppose the results of the work of the field teleologist. Both Davidson and Quine take for granted that their investigators can recognize a linguistic agent—and thus an agent—when they see one. But the field teleologist can't presuppose even the ability to recognize the difference between agents and nonagents. In the relevant sense, an agent is an entity that acts, and one can have reason to believe that an entity is acting only if its behavior can be interpreted to have some point. So if it is unknown whether there are any goals in the neighborhood, it is also unknown whether there are any agents in the vicinity. Agency and goal-directedness are related notions, and the field teleologist must identify them together or not at all.

But what questions, exactly, must the teleologist answer? The exploration party wants an answer to a question that might turn out to have practical consequences: "Is there any behavior or activity around here that has a goal, and, if so, what is that goal?" This question has practical importance because a goal attribution involves more than mere description. It also involves an explanation of events that have that goal. And this explanation has implications for what we can expect in the future.

## 2. GOAL-DIRECTEDNESS: DESCRIPTION AND EXPLANATION

Consider once again the biological description of the *Sphex* wasp that I quoted in the introduction. "When the time comes for egg laying, the wasp *Sphex* builds a burrow for the purpose and seeks out a cricket which she stings in such a way as to paralyze but not kill it. She drags the cricket into the burrow, lays her eggs alongside, closes the burrow, then flies away, never to return. In due course, the eggs hatch and the wasp grubs feed off the paralyzed cricket, which has not decayed, having been kept in the wasp equivalent of deep freeze." Wooldridge ultimately comes to the conclusion that there is no logic or thoughtfulness involved in what the wasp does, but he never takes back the suggestion that she builds the burrow "for the purpose" of laying her eggs or the assertion that she "seeks out" the cricket or the implication that the stinging of the cricket has the goal of feeding her growing young. The behavior of the wasp has a goal, even if it is not the result of any rational thought on the part of the wasp. So causation by the beliefs and desires of the agent can't be part of what it is for the behavior to have a goal.

Notice that Wooldridge in no way suggests that what the wasp does could not be explained in entirely physical terms. But he makes no attempt to do so. Rather, in describing the behavior of the wasp, he helps himself to an especially useful shortcut, the teleological locution, which also has the additional benefit of providing the rudiments of an explanation for the behavior. Instead of giving a detailed physical description of the motions of the wasp prior to her hunting foray, he simply says that what she does is "build a burrow." (Notice also that the wasp could be building a burrow even if what she does is, for contingent reasons, not entirely successful at bringing about a completed burrow.) But if she is building a burrow, then we can explain this act by appealing to the supposition that what she does has the further goal of laying her eggs (in an optimal place).

The field teleologist must determine whether there is anything happening on Sirius that is significantly similar to the wasp's building burrows, seeking out crickets, and preparing for the laying of her eggs. But what, then, is involved in the claim that the wasp builds the burrow for the purpose of laying her eggs inside? On the one hand, such a statement describes the motions of the wasp in a certain way. On the other, this descriptive content implicitly embodies an explanation of the behavior that is being described. The teleologist must ascertain both whether anything is happening in her environment that satisfies the descriptive constraints on a behavior's having a goal and whether this "having a goal" helps to explain the behavior. The descriptive and explanatory aspects of goal-directed language are intimately linked. But what is the nature of those linkages?

Let's start with the descriptive component of the goal-directed ascription to the wasp's motions. In fact, the attribution describes those motions in two related ways: what the wasp does is "build the burrow," and this preparation (which, remember, is also the set of motions that the wasp carries out) is itself "for the purpose of laying her eggs." Both phrases describe the wasp's motions in teleological terms as being for the achievement of some goal. In the case of the description of the motions as "for the purpose of laying her eggs," the teleological character of the description is explicit. But the description of the behavior as "'building the burrow" is equally a description in terms of a goal. After all, that the burrow should come to exist is the point of the wasp's motions, and even if the burrow never comes to be (a malevolent scientist keeps undoing the wasp's work), what the wasp does is still "building the burrow" because that is what the frantic motions are *for.*

These two teleological descriptions of the same behavior are nested, to make a very poor pun. As the biologist says, the wasp builds a burrow *for the purpose* of laying her eggs. The frantic motions of the wasp are both for building a burrow and for laying her eggs, and furthermore the building of the burrow is itself for the laying of the eggs. So the same behavior has two goals, one proximate and the other distant in time from the behavior itself, and it is appropriate to say that the behavior described in terms of the proximate goal is for the goal of realizing the distant end. This multiplicity of goals for a single event or behavior is typical of goal attribution. The broken wing display is both for the purpose of deceiving the intruder and for the protection of the plover's young, and the plover deceives the predator in order to protect her young. Agents attempt to do one thing by attempting to do another thing first.

It is not accidental that this is the case. If the wasp's motions that we describe as "building a burrow" were not preparatory to the laying of eggs or to some other further end, we would be far less likely to think that what was being done was the building of a burrow. There are all sorts of natural processes that frequently result in formations similar to wasp burrows. But these processes, and the "burrows" they produce, seem to have no point. That is, they don't seem to be for anything. In general, it is far more likely that some behavior is for some goal when that goal itself can be correctly described as serving some further end. Because the goal is not behavior that can be for anything (neither the burrow's having been built nor the intruder's having been deceived is a behavior of the agent), we tend to redescribe some actual behavior of the agent as acts of burrow building or predator deception, which themselves are then described as being for some further end. Since descriptions in terms of goals involve this systematic fitting together of proximate and distant goals, the same behavior tends to accrue a series of nested, goal-laden descriptions: what the wasp does now is *(an attempt) to build a burrow for providing a suitable place for the laying of eggs.*

The state of affairs that is the goal of the behavior is not typically identical with the behavior that has that goal. The goal of the wasp's behavior is for the burrow to have been built or for the eggs to have been laid in the burrow, but the motions themselves are not a built burrow or laid eggs. The one event, the motion, is described in terms that refer to the other state of affairs, the built burrow, but the burrow is not identical with the motions of the wasp that may result in the burrow. Such descriptions of one event in terms of another are quite common, and acts are often described in terms of events they cause to occur.[3] In the case of goal-directed descriptions, however, the state of affairs referred to in the description need not be caused by the events that are being described, although they might be. The wasp can be building a burrow even if she fails to achieve this end (because of experimental interference, for example), just as I can be walking to the store even if I never get there. In such cases, we often introduce the phrase "attempting to" or "trying to" to signal the chance of failure, as when we say that the wasp is attempting to build the burrow. But even in cases of failure, what the wasp does is just as much *for* this preparation as when she succeeds in her goal.

Events that are for some goal must be pretty good at reaching that goal. The standard cases of goal-directed behavior involve systems that are working effectively. The homeostatic processes that we think of as having the goal of maintaining a constant temperature in warm-blooded animals are pretty good at keeping that temperature relatively stable. The photosynthetic processes in plants that seem to be for nutrition are pretty good at providing nutrition. Unless hunting behavior is pretty successful, we don't think to call it hunting. We would never think of attributing a goal to the behavior of a system unless that system was pretty good at reaching that goal. That is, if something is for reaching a goal, it must be pretty successful at reaching that goal.

The question of how we should articulate this insight, however, remains quite open. And this is an especially hard insight to get right. The link can't be causal. As we just saw, it is central to the logic of goal attributions that the fact that B is for G does not require that G should occur. So it can't be necessary that B cause G in order for B to be for G. Nor can it be sufficient. Many events cause others without being for that result. Nevertheless, that B is for G has *something* to do with B's being appropriate for bringing about G, and being appropriate for bringing about G has *something* to do with actually producing G. If what the wasp did were unrelated to bringing into existence a burrow in which she then laid her eggs, then we would have no more reason to think that the motions were for the preparation of a burrow than we would have to assert that they are for flying to the moon. But one must be very careful in specifying this link between a behavior and its goal.

For one thing, the fact that B can be for G without causing G is indicative of a more general fact about the logical status of the goals of behavior. The event or state of affairs mentioned in the description as the goal of some behavior is not itself a particular event or state of affairs at all. When the sting of the wasp causes the paralysis of the cricket, both the stinging event and the state of affairs that is the paralysis of the cricket must be actual. The causal relation is a real relation between real particulars, just as the "taller than" relation is. But when the wasp builds a burrow for the sake of laying her eggs, there is no guarantee that she will ever lay them. Since it is hard to see how there can be a real relation between an actual particular and a merely possible state of affairs, it would seem to be impossible that the building of the burrow's being for egg laying could involve any real relation between the building and any actual, particular act of egg laying.

Another possible articulation of the sense in which an act's being for the achievement of some goal implies that the act is appropriate for bringing about that end is the suggestion that this "bringing about" can be interpreted in a statistical way. Some particular behavior—for example, the building of a burrow—is for some end—for example, the laying of eggs in the burrow by the burrow maker—only if the behavior tends to bring about the end in the sense that it makes the end more likely. This suggestion, borrowed from Larry Wright's attempt to articulate the descriptive content of goal attributions,[4] has the merit of not implying that goal ascriptions involve the attribution of a fictitious real relation between the behavior and the goal. When one says that building the burrow makes egg laying in the area more likely, all one need be asserting is that it is more likely that *some* event that satisfies the description "egg laying in this area" will occur if the burrow is successfully built than if it isn't. Similarly, that these motions are for the building of a burrow implies only that it is more likely that there will be *some* burrow built if the wasp moves in this way than if she doesn't, not that these motions in fact will bring about a particular burrow. In this view, the goal of a particular behavior is not itself a particular but rather a *type* of state of affairs, and the behavior increases the likelihood that some token of this type occur.

As we will see in the next section, this suggestion doesn't quite pick out the descriptive content of goal ascriptions. But it does have the virtues of recognizing both that a behavior's being for some goal does not imply a real relation between the behavior and the goal treated as particulars and that whatever it means to say that B is for G, it involves some modal or statistical variation on B bringing about G.

As Aristotle pointed out, one of the ways in which we ask "why" involves expecting an answer in terms of a goal. If I ask my student why she is sitting in class, the type of answer that I expect would specify the purpose of her being there. Any other type of answer, in terms of what it is for her to sit there, for example, or in terms of the bodily motions that brought her to this position, would be unsatisfactory. Similarly, when one is informed that the wasp is moving in this way in order to build a burrow, this information provides an answer to why the wasp acts in this way, in one of the senses of this question. As such, goal attributions perform an explanatory function as well as a descriptive one.

Every explanation involves an explanandum—something to be explained—and an explanans—something that provides that explanation. In many kinds of explanation, what is explained is that some particular event occurs or occurred: this seems to be the case in goal-directed explanations. What is unclear is just how goal-directed attributions manage to explain the event that is described as having a goal.

One common kind of explanation, the deductive-nomological kind, explains an event by describing it as an instance of a type of event and then going on to note that an event of this type can be expected to occur after some other event of another type has occurred and that an event of this other type has in fact preceded the event in question. Question: Why does this liquid bubble? Answer: This bubbling is a case of water boiling. Water boils when sufficient heat is applied to it, and heat has been applied to this water. At the least, a deductive-nomological explanation of the occurrence of an event demands a perspicuous description of the event to be explained, a perspicuous description of another event that preceded the event to be explained, and a universal generalization or covering law linking events of the types mentioned. This type of explanation is nomological in virtue of its appeal to laws; it is deductive because the explanandum is treated as a sentence that is deduced from the two sentences composing the explanans. The fact that the explanation is provided by a logical relation among sentences guarantees that deductive-nomological explanations are intensional: the description of the event explained and the events relevant to the explanation determine whether an explanation has been provided.

On the surface, goal-directed explanations do not fit into the deductive-nomological pattern. The *only* event referred to in the explanation is the event to be explained, and it is referred to twice. The motions of the wasp are for building the burrow, and this explains why these motions occur; but the building of the burrow is the same event as the motions of the wasp.[5] Even if one waives the requirement that two different events be referred to in the explanans, there is no appeal to a covering law in the explanation. The only plausible candidate for such a law would be something like: "Whenever there is a building of a burrow by a wasp, there are these motions." But, as Carl Hempel's classic discussions of teleological explanation make clear, what in particular an agent does to reach its goal is rarely the only thing that it does to reach its goal, and it is never the only thing that it could do to reach it. So it seems false that a wasp's moving its wings in this way is a necessary condition on building a burrow.[6]

We sometimes give a causal explanation for the occurrences of some event instead of offering a deductive-nomological explanation. In such cases, we explain why the event occurs by specifying a preceding event that caused or brought about the event to be explained. Sometimes causal explanations are accompanied by a causal law that backs up the claim that the first event caused the second. But this is not necessary. Sometimes we know that some event caused another without knowing any law linking the two. Because such explanations assert a causal relationship between individuals rather than a logical relation among sentences, such explanations are extensional. If event A causes event B, and in virtue of that fact A's occurring explains B's occurring, then how A and B are described is irrelevant to the truth and adequacy of the explanation.

Goal-directed explanations are not causal in character. They refer to only a single event and, furthermore, are intensional rather than extensional in character. It is precisely the fact that these motions are correctly described as being "for the building of the burrow" that explains why they occur. The fact that these same motions are also correctly described as "the only motions of the wasp's limbs that occur at time t and cause sound waves" is irrelevant to the explanation offered. So the sentence "the motions are for the building of the burrow" is true, and the sentence "the motions are for the production of sound waves" is false, even though the motions that are the building of the burrow are also the motions that are the causing of the sound waves.

There is another respect in which goal-directed teleological explanations differ from both causal explanations and deductive-nomological explanations. It has often been noted that explanation and prediction are twins; when one has the one, one also has the other. This is as true of explanations of what behavior is for as it is of other types of explanation. But *what* such explanations allow us to predict is very different from what the other types of explanations allow us to predict. In general, explanations of either the causal or the deductive-nomological kind allow one to predict that when the same type of circumstances prevails in the future as prevailed at the time of the event explained, the same type of event will occur. Explanations in terms of goals are mostly silent on such matters. Instead, what one can predict when one has a goal-directed teleological explanation is future behavior by the agent of the previous behavior in different circumstances. If one knows that the building of the burrow by the wasp is for the laying of eggs, one also knows that the wasp will do other things for that same end as her situation changes. This allows one to predict, in a general sort of way, what type of behavior to expect from the wasp in the future. More important, knowing that the lion is crouching for the sake of concealment and that this concealment, in turn, is for sneaking up on animals so that they can be killed, allows one to predict the lion's future behavior and to take appropriate countermeasures. But this is a very different type of prediction from that provided by standard causal or deductive-nomological explanations.

Despite these obvious differences between goal-directed and causal explanations, some have insisted that, insofar as goal-directed explanations are legitimate, they must be causal. There are two sources of this conviction. First, there is the assumption that *all* legitimate explanations are causal: so if goal-directed explanations are legitimate, they must be causal. What stands behind this assurance is probably the conviction, derived from the development of physical science over the last three centuries, that there is a causal explanation for everything that occurs and that these causal explanations render all other types of explanation redundant or otiose. This tendency to think in terms of causal explanation is reinforced by a second consideration, the fact that *rational* explanations, which clearly share many features with goal-directed explanations, can easily be made to fit into a causal model. As I have insisted, the only other "state" or "event" that is mentioned in a goal-directed explanation, the goal itself, is no event at all. Were such an event to occur, it could occur only after the event to be explained. So the goal can't *cause* the event to be explained. But the rational agent's *thoughts* regarding the goal are genuine events and do precede the event to be explained, so these thoughts are available as potential causes of that event. Since it is assumed that all real explanations are causal, it seems to follow that all goal-directed explanations are implicitly causal explanations appealing to the hypothetical beliefs and desires of a rational agent.

Such a position, however, ignores the evident structure of goal-directed explanations, which makes no reference to the beliefs and desires of agents. It also ignores the ways in which we use such explanations to gain insight into the acts of agents that we do not suspect of rationality. And, while the general point regarding the ubiquity of causal explanation must be handled with care (and will be discussed in due course), it need not detain us now. After all, our example has stipulated that the field teleologist already knows that everything that happens on Sirius can, in principle, be explained in causal terms. So, for now, we will put aside such causalist reconstructions of goal-directedness.

It seems, then, that goal-directed explanations differ from other types of explanation in several significant respects. First, goal-directed explanations cite only one event, the event to be explained, rather than two or more. Second, goal-directed explanations differ from causal explanations in that they are intensional and from deductive-nomological explanations in that they don't seem to appeal to any laws. Finally, goal-directed explanations help us predict the future behavior of an agent when circumstances differ from those that prevailed at the time of the event explained rather than to predict what will happen when the same circumstances prevail in the future.

The evident link among these peculiarities of goal-directed explanations is that the behavior explained is seen as occurring *because* it satisfies the description of being for the goal that is cited. It is the description of the behavior as being for its goal that provides the explanation. Therefore, only one event need be referred to, the event to be explained. Because it is the redescription that provides the explanation, it is the particular description contained in the explanans, rather than the referent of that description, that is crucial to the explanation. And, as we shall see, since the explanation by way of redescription involves placing the event to be explained in an ongoing context of behavior, it provides the basis for predicting the future behavior of the agent.

This discussion of the descriptive and explanatory aspects of goal-directed explanations leads to the following conclusions. If some event happens for some purpose, then the event occurs *because* it is for that purpose. This is the lesson of our discussion of the explanatory aspect of goal attributions. And what it is for an event to be for a goal involves, in some way, that event's being appropriate to or tending toward the realization of the state of affairs mentioned in the goal. This is the lesson of our discussion of the descriptive aspect of goal attributions. Two crucial points remain entirely obscure, however. Just how are we to understand the sense in which an event that has a goal "is appropriate to or tends toward" bringing about that goal? And just how are we to understand the sense of "because" in which an event happens because it is for its goal?

The two conclusions of this section form the starting points for Larry Wright's revolutionary account of teleological language. In effect, he presents concrete suggestions for how to interpret the sense in which a behavior that has a goal is appropriate for bringing about that goal and the sense in which an event that has a goal happens because it has that goal.

## 3. WRIGHT'S ANALYSIS OF GOALS

If all goal-directed behavior is rational, then it is easy to understand the sense in which an event that has a goal is appropriate to bringing about that goal. Rational behavior is explained by the beliefs and desires of the agent. The behavior, insofar as it is rational, is performed for a reason. In the simplest case, that reason is supplied by the contents of the beliefs and desires that explain the behavior. The behavior is rational only if it will bring about the desired end, provided that the beliefs on which the behavior is based are true. Flipping the switch will result in the light's being on—which is what I desire—if what I believe regarding light switches is true. That is, rational behavior is appropriate to bring about its goal in a situation in which the agent's beliefs are true. In addition, if some version of the principle of charity is accepted, and most of a rational agent's beliefs must be true, then, if we hold that all goal-directed behavior is rational, we can also understand the sense in which goal-directed behavior tends to bring about its end. If rational behavior is successful when the agent's beliefs are true, and if those beliefs are true most of the time, then most of what a goal-directed agent does to achieve a goal must be successful in attaining that goal.

If all goal-directed behavior is rational, it is also easy to understand the sense in which goal-directed behavior happens because it is appropriate for achieving its end. Since it is the beliefs and desires of agents that explain rational acts, and since rational agents—insofar as they are rational—act because, given their beliefs, what they do is likely to bring about what they desire, what causes a rational agent to perform a given act is precisely its appropriateness to achieving the goal, given that the agent's beliefs are true. That is, the feature of the act that accounts for its occurring is that it is appropriate to the ends of the act.

Because the hypothesis that the agent of goal-directed acts is rational renders the structure of goal-directedness so transparent, it makes the suggestion that all goal-directed behavior is the product of the mental events of rational agents quite attractive. In this view, any claim that some particular piece of behavior has a goal involves a tacit reference to what Davidson calls a teleological theory of the behavior of the agent.[7] Such a theory attributes both goals to the behavior and beliefs and desires to the agent at one blow, consistent with the agent's actual behavior and under the constraints of rationality. In such a mentalistic conception of teleology, beliefs and desires are theoretical entities that are introduced to explain goal-directed behavior, but the description of the behavior as goal-directed is dependent on the success of the theoretical rational structure in explaining the details of the behavior. So in this view there is no teleology without mental events that have content.

Standing against this view, which has been dominant at least since the mid-1960s, is the intuition that some behavior has a goal even though it lacks rationality. The reproductive routine of the wasp would seem to be an exemplary case of such goal-laden but arational behavior. One who is committed to the mentalistic analysis of goals may attempt to account for such cases by appealing to the notion that the attribution of goals to the behavior of the wasp is metaphorical and based on similarity between the wasp's behavior and rational behavior, although this suggestion is not entirely satisfactory. Thus it is incumbent on those who think that behavior can have a goal even when its agent is not rational to articulate an alternative sense in which goal-directed behavior is appropriate to its goal—a sense that does not appeal to the beliefs of an agent—and a sense in which the appropriateness of the behavior to achieve that goal explains why the behavior occurs—a sense that does not appeal to the rationality of the agent.

Larry Wright's account of goal-directedness tries to respond to this challenge of making sense of both the appropriateness of goal-directed behavior and the explanatory character of goal-directed descriptions without invoking hidden mentalistic causes. The key to the strategy is a statistical or probabilistic notion of appropriateness. As noted above, events that are for some goal must be pretty good at reaching that goal, and the standard cases of goal-directed behavior all involve systems that are working effectively. According to Wright:

> (T) S does B for the sake of G iff:
>
> (i) B tends to bring about G
>
> (ii) B occurs because (i.e., is brought about by the fact that) it tends to bring about G.[8]

Wright's intuition here is plain enough. Consider the behavior of the wasp preparatory to the laying of her eggs. In very many cases, this behavior results in healthy and well-fed wasp grubs. This overall result would not be so frequent if each of the stages along the wasp's life way did not frequently result in the ends we attribute to them, such as "building a burrow." If this weren't the case, there wouldn't be so many wasps. So by acting as they do, wasps, in a rough and ready sense, tend to bring about built burrows and healthy young wasps. Hence, Wright's clause (i). But why does the wasp act in *that* way? Because it tends to get the burrow built, of course. So the feature that brings about the wasp's motions is the fact that this sort of motion gets the burrow built, and the feature that brings about the motion that gets the burrow built is the fact that burrow building, in turn, when combined with other acts of the wasp, tends to bring about healthy young. Hence, Wright's clause (ii).

Notice that neither of these clauses tacitly refers to intentional mental events. Condition (i) seems to be a straightforward statistical claim, and condition (ii) mentions only a feature of the event that is to be explained, that *it* tends to bring about G. It does not refer to any other, mental, events.

Wright claims that if his formula is right, then the goal-directedness of a given event is testable. If one varies the environmental conditions, then different events will lead to the same goal, G. If the behavior of the agent varies in such a way that for every change in environmental conditions, the likely outcome of the behavior remains G, then we have reason to believe that the entire sequence of behaviors has been brought about by the fact that each of the behaviors tends to bring about G. This, in turn, confirms that the entire sequence is for the sake of G.

That anyone attempting to apply Wright's formula must appeal to this sort of evidence helps to clarify the character of his formula. In Wright's view, the claim that an event has a goal contains a counterfactual as well as a realistic component. On the one hand, clause (i) asserts that, in the actual circumstances, what occurs tends to bring about goal G. On the other hand, formula T suggests that, were the actual circumstances different, something else would occur and that what occurred would tend to bring about G in those circumstances. It is this counterfactual element that fleshes out the sense that we should give to the phrase "B is brought about by the fact that it tends to bring about G" in clause (ii).

Although Wright is mostly silent on the subject, the character of the evidence that is relevant to forming judgments of goal-directedness also helps to clarify the role of the agent S in formula T. The reference to the agent in Wright's formula is not vacuous. Behavior is constituted by events or sets of events, and events are alterations in *objects* (or, if Jaegwon Kim is right, properties of objects at a time). The fact that the formula makes reference to a particular agent S reflects the fact that it is not enough for B to be for G, for B to tend to bring about G, and for *something* to happen in other circumstances that would tend to bring about G in those different circumstances. The fact that both Wright and our ordinary ways of speaking refer to an agent indicates that what happens in those altered circumstances must be alterations *in the same object* as the alteration that is being explained teleologically. What is relevant to determining whether what the wasp does is for the goal of building a burrow is not what the experimenter would do in different circumstances: it is what *the wasp* would do. That is, goal-directed attributions implicitly refer to agents, as indicated by the sort of evidence that is relevant to making such attributions.

Wright's formula T handles the problem of specifying the sense in which goal-directed behavior is appropriate to its goal in a way very different from that in which a mentalistic approach does so. If all goal-directed behavior is rational, then a goal-directed act is appropriate to its goal in the sense that it would result in the achievement of that goal if the beliefs of the agent that motivate that act were true. But Wright's formula T does not demand that the agent have any beliefs at all. Instead, it suggests that an act is appropriate to its goal only if it tends to bring about that goal. Similarly, Wright's formula differs from mentalistic interpretation in its handling of the question of how we are to understand the explanatory force of goal-directed language. For the mentalist, to say that an act occurs because it is for a goal G is just an elliptical way of saying that the act is caused by states of the agent that have the structure of being a desire for G and a belief that G will occur in the current situation if the agent performs the act. In Wright's formula, there is no mention of the content of the states of the agent. Instead, there is only a reference to the fact that this behavior tends to bring about G and a tacit reference to the fact that, were the circumstances different, the behavior would also differ in such a way as to tend to bring about G. The formula is entirely silent on why this is so—that is, why the behavior has the goal it does—whereas the mentalist explains this point, as well as the fact that the behavior occurs, by appealing to the desires of the agent.

Were Wright's formula adequate, it would be a great help to the field teleologist. She would formulate the hypothesis that some sequence of events, all of which tended to bring about result G, was for goal G. She could then vary the environment and see whether some new sequence of events occurred that also had the property of tending to result in G. If those sorts of events did occur, she would have reason to believe that the whole expanded sequence was for the sake of G. If there were any entity in that environment that seemed to change its properties every time the environment were altered, she would have reason to think of that entity as an agent and to think of those changes as the agent's basic acts. And, were Wright's formula correct, she would be able to do all of this without any prior knowledge of which entities, if any, were alive and which entities, if any, had mental states, and without needing to make any assertions about the mental causes, if any, of the events she is observing.

Unfortunately for the teleologist, however, there are two respects in which Wright's proposals are unacceptable. First, the view falls into serious type/token confusions having to do with the logic of tendency statements, confusions that seriously undercut the claim that B is for the sake of G only if B usually brings about G. Second, and equally damaging, it is incoherent or vacuous to assert, as clause (ii) does, that B is brought about by the fact that it tends to bring about G. Taken together, these problems are deadly to Wright's approach. We will discuss each of them in turn.

It is pretty clear that Wright means his B's, the actual behaviors that tend to bring about the goal, to be individual events. If he didn't, his proposed test for verifying that some event is brought about by the fact that it tends to bring about a goal would make little sense. Perhaps more important, in all ordinary contexts in which we explain an event by appealing to its goal, what we explain—the B—is an individual bit of behavior, not a type of behavior. The wasp does *this* because she is building a burrow. So the claim that event B is for G only if B tends to bring about G must apply the tendency to the individual event. For clause (i) to make logical sense, the occurrence of B must tend to bring about G. On the other hand, as we saw above, the goal of the behavior, G, can't be an individual event but must refer to a type of event.

The tendency language in clause (i) seems to have a statistical sense. Wright is well aware that the condition "B brings about G" is too strong to articulate the sense in which B must be appropriate to bringing about G if B is to be for G. In teleological language, of course, a behavior may be for some end that it fails to bring about. So Wright advances the weaker condition that B must tend to bring about G, where "tendency" suggests a statistical rather than a strictly causal relation.

Tendency language usually refers to the statistical probability that some single event will have some specific outcome when a large number of similar events that occur in relevantly similar circumstances have that same specific outcome a given percentage of the time. If there is a greater than 50 percent probability that G will occur if B occurs (in the sense that events similar to B that occur in circumstances similar to those in which B occurs bring about this result in better than 50 percent of the cases), we say that B tends to bring about G. For example, if, in normal conditions, I were to flip a coin that is more heavily weighted on the tails side, there is a greater than 50 percent chance that the coin will land face side up; in that sense, this flip will tend to bring about an instance of the coin lying face side up.

There is a second sense in which we use tendency language. There can be a greater probability that G will occur if B occurs in a given situation than if it doesn't occur, when the probabilities of both outcomes are measured as in the first case, above. So, once again in normal conditions, there is a vanishingly small probability that any given sperm cell swimming up a vagina will result in an egg's being fertilized by that sperm.

Nevertheless, such swimming might tend to bring about an event of the fertilizing type in the sense that this outcome is more likely to occur if the swimming takes place than if it doesn't.

Both of these senses in which an individual event might tend to bring about a type of result depend on our ability to assign a probability of occurrence to a type of result given the occurrence of that individual event. There are two senses in which we can do this. Individual events are *individual.* They always occur in a particular determinate context. What results from this occurrence in this particular context might not be causally fixed by the conjunction of circumstances and event. This is certainly true in quantum cases, and, for all I know, it is true in the case of the coin flip: there is only a certain probability that this individual event will result in a given type of outcome. Alternatively, what results from a particular occurrence in a particular context might be entirely causally fixed by the conjunction of circumstances and event, but we might not be in a position to know this: there might be an inductive or Carnapian probability that the result occurs, which really represents epistemic probability.

There are four ways to interpret Wright's tendency language that are compatible with the logical structure of goal-directed judgments, that is, with the fact that the behavior referred to in the judgment is individual while the goal is not. The tendency can be either absolute or relative to what would be likely to happen were the event not to occur, and the probabilities in terms of which the judgment is made can be epistemic or causally nondeterministic. But *none* of these interpretations gives us an acceptable interpretation of the sense in which goal-directed behavior must be appropriate to bringing about its goal.

Very frequently, when goal-directed behavior fails to achieve its goal, it fails because the agent's act is the wrong thing to do to bring about the end in the *actual* circumstances. When this occurs, none of the senses of "tendency" that might be acceptable interpretations of Wright's tendency talk are at all relevant. This is obviously true in cases of rational behavior. If I try to replace an English bayonet lightbulb by screwing the new bulb into the socket, what I do is for connecting the bulb with a power source. But what I do fails because it is the wrong thing to do in the *actual* circumstances. In the actual circumstances, what I do does not tend to bring about the end in any of the four possible senses of "tend." There is neither an epistemic nor real probability greater than 50 percent that

I will get light by doing this, nor is there a greater probability in either of these senses that I will get light if I do this than if I don't. We mark this sort of error in action by saying that my action fails because I have false beliefs regarding how English lightbulbs are joined to lamps. But we mark the fact that the act is nevertheless for getting light by saying that I desire to replace the bulb.

This same pattern of error in behavior in the actual circumstances is routine in failures of nonrational goal-directed behavior. If there is no viable egg present, nothing that the sperm cell actually does in any way increases the likelihood of *that* sperm fertilizing some egg, but its undulations, nevertheless, are for such fertilization. When a plover deceives a predator into following her away from the nest rather than searching for the plover's eggs, what the predator does is for getting a hearty meal. What it does in the actual conditions, however, *decreases* the probability of reaching this end, in the only senses of probability that apply. Similarly, nothing that a wasp does to build a burrow tends to increase the likelihood of the burrow's getting built if the actual materials she is trying to use are entirely unsuitable.

The problem here is obvious. In all of these cases, it is true that *instances* of the *type* of behavior that the agent engages in are likely in *normal* circumstances to tend to bring about the end, in one or another of the senses of "tend." But these normal circumstances need not be actual, nor need they be statistically probable. So it is simply not true that every individual event that is for some goal tends to bring about that type of goal, in any acceptable sense of "tends." Nor is it even true that instances of a type of behavior tend to bring about their goal, in any of the senses of "tend," more often than not. This result implies that Wright's clause (i) simply fails to capture the sense in which goal-directed behavior must be appropriate to its goal.[9]

Wright's clause (ii)—B is for the sake of G only if B occurs because (i.e., is brought about by the fact that) it tends to bring about G—attempts to capture the sense in which goal-directed ascriptions are explanatory. Clause (ii) piggybacks on clause (i). So if there are serious problems with clause (i), as we have seen that there are, those problems are inherited by clause (ii). Nevertheless, clause (ii) exemplifies a strategy for attempting to articulate the explanatory structure of goal-directed descriptions that is independent of the way in which we understand the descriptive content of those descriptions. For that reason, it is worthwhile to consider clause (ii) in its own right, regardless of its dependence on the inadequate clause (i).

Behind Wright's strategy for interpreting the explanatory power of teleological explanations is his insight that, in cases in which an agent acts for a goal, the agent most often would do something else that would be appropriate to reach that goal were the circumstances of the act different. When the intruder moves toward the plover, the plover goes into the broken wing display, which tends, in a rough and ready sense, to induce the predator to follow her. But if the predator ignores the plover's apparent distress, she responds by moving toward the intruder, thus making herself more conspicuous, which in *those* circumstances tends to induce the predator to follow her. So, Wright reasons, the plover acts as she does in the actual circumstances because doing what she does tends to induce the intruder to follow her. This judgment is warranted by the fact that the plover, in different circumstances, would act in a way that in those circumstances would tend to result in the intruder's following her.

This entire discussion would seem to be perspicuous. The issue becomes complicated, however, when we try to articulate the exact nature of the explanation suggested by Wright's clause (ii). For Wright, evidence of goal-directedness amounts to a correlation between what an agent does in varying circumstances and what event would, in those circumstances, tend to bring about the goal that is attributed to the act. This correlation suggests that we should read the "because" in clause (ii) in a deductive-nomological fashion. On this reading, what the agent does in an actual situation would be explained by our inferring a sentence stating what she does from a statement asserting some general law and a statement asserting an instance of the antecedent of that law. The problem with this reading, however, is that there doesn't seem to be any law in the area that could occupy the appropriate role in the inference. All the potential acts of the agent share a tendency to bring about the end in the prevailing circumstances, whatever those prevailing circumstances are. But the agent, of course, will not *always* act in a way that will tend to bring about that end. So there is no true law to the effect that the agent will always act in a way that will tend to bring about this end. Rather, we can expect that the agent will act in a way that will tend to bring about this end only if *its current acts have this end*—that is, if it currently desires this end or, if we omit the mentalistic reference, if the agent is acting for the sake of achieving this goal. So, if we read clause (ii) in a deductive-nomological fashion, the full expansion of the explanation turns out to be:

> (1) If agent S is acting for the sake of goal G, then it will do something that tends, in the circumstances, to bring about G;
>
> (2) S is acting for the sake of G in circumstances C;
>
> Therefore (3) S does something that tends to bring about G in circumstances C, that is, B.

The problem with this deductive-nomological interpretation of Wright's position is evident from this expansion. Clause (ii) occurs as part of a definition of what it is for S to act for the sake of G. So any interpretation of clause (ii) that takes for granted what it is for S to act for the sake of G, as this one does, creates a vicious circle. So the "because" in Wright's clause (ii) can't be interpreted in deductive-nomological fashion.[10]

Wright recognizes this circularity and therefore introduces the parenthetical phrase in clause (ii). We are to understand the "because" in the clause as "brought about by the fact that" and so are to read the entire clause as "B is for the sake of G only if B is brought about by the fact that it tends to bring about G." This suggests that Wright means the explanation here to be causal, and his other comments confirm this reading. But if interpreted this way, the clause makes no sense. It could be read either as a singular causal statement or as a compressed causal explanation. Assertions of causal connection assert relations between two (or more) individuals: the event that is caused and the event that is the cause. The cue ball's being struck ($e_1$) caused its motion ($e_2$). Roughly, what is asserted is that the occurrence of $e_1$ brings about $e_2$. But in goal-directed descriptions, only the event to be explained is mentioned, even if Wright's analysis suggests that it should be mentioned twice under different designations. (I leave out the goal, which, as we have seen, is no individual at all.) The event with the name "B" is brought about by the event that is such as to tend to bring about G, an event that happens to be the same event as B. That is, if Wright is correct, B causes B. But absent some possible acts of God, such a self-relation makes no sense.

Alternatively, Wright could be suggesting that B's tending to bring about G is part of a causal explanation. In causal explanations, alluding to one of the correct descriptions of an event is part of the explanation of that very event. If there is a law relating events of the type "cue ball being struck" with events of the type "cue ball moving," then we can (elliptically) explain the occurrence of *this* event—call it "B"—by saying that "B is an example of a cue ball moving." One can do this, however, only if the ellipsis includes two vital pieces of information: the law of which "cue ball moving" is the consequent and the statement of the fact that the antecedent of the law ("cue ball struck") is satisfied. But Wright seems to think that *just* the fact that B tends to bring about G can causally explain B. At least he offers us no clue as to what laws might be lurking in the area. Such a causal explanation is neither causal nor an explanation.[11]

Wright's formula T articulates neither the manner in which a goal-directed explanation explains the behavior that has the goal nor the descriptive content of goal-directed discourse. Nevertheless, the nature of these failures is instructive. Wright fails not because his intuitions are flawed, but because he has not adequately articulated the force of those intuitions. The intuition behind his attempt to understand the descriptive content of goal-directed descriptions is that if the behavior of an agent has a goal, then it must be appropriate to bringing about that goal and that if the behavior is appropriate for bringing about that goal, then the behavior of the agent must be successful at achieving that goal a substantial portion of the time. In the next chapter I offer another attempt to capture the force of this intuition, one that differs fundamentally from Wright's in that it distributes this substantial degree of success over the totality of the agent's behavior rather than restricting it to a particular type of behavior as Wright's does. The intuition behind Wright's attempt to articulate the explanatory force of goal-directed attributions is that goal-directed behavior happens because it is appropriate to bring about the agent's goal. Once we have correctly understood what it is for an agent's behavior to be appropriate to its goal, we will be able to correctly understand how the fact that some behavior is appropriate to a goal helps us to explain that behavior.

CHAPTER THREE

# THE DESCRIPTIVE CONTENT OF GOAL ATTRIBUTIONS

Larry Wright's account of goal attributions is based on two crucial insights. First, Wright recognizes that when one says that some agent performs an act for the sake of attaining a particular goal, there is some sense in which what the agent does is appropriate to that goal. Second, Wright recognizes that the goal-directed event occurs *because* it is appropriate to its goal, and, for that reason, citing the fact that some event is for the sake of attaining a goal can explain the occurrence of that event. Nevertheless, Wright does not offer an adequate account of the sense of goal attributions. As we saw in chapter 2, there are two central problems with Wright's account. First, he does not adequately articulate what is contained in the fact that what is done for a goal is appropriate to attaining that goal. Second, his causal analysis of the explanatory power of goal-directed explanations is flawed.

In this chapter, I offer a revised articulation of Wright's insight that goal-directed behavior is appropriate to attaining its end. In light of this discussion, in the next chapter I offer a revised understanding of the explanatory character of goal-directed explanations.

## 1. THE SCOPE OF THE JUDGMENT OF APPROPRIATENESS

There are several senses in which a certain event or behavior can be appropriate to attaining a given end. In the simplest possible case, any act that brings about a particular result is appropriate to attaining that result. When an engineer is attempting to design a system to attain a given result, he needs to figure out what series of changes in the current environment will bring about that new state of affairs. And if his design accomplishes that series of changes, then what his creation does is appropriate to its goal.[1] But, as Wright recognizes, not all goal-directed behavior is appropriate to its goal in this sense: a behavior can have a goal even if it does not result in the attainment of that goal.

A behavior is also appropriate to a goal if, in the current circumstances, it could rationally be expected to bring about that goal, even though it doesn't. That is, an event that has a high statistical probability of achieving its goal in a given situation is the appropriate thing to do in that situation, even if success is not guaranteed. It is this sense of appropriateness that Wright accepts in his account of goals.

The problem with Wright's statistical analysis is that it fails to take into account one of the common ways in which a behavior can have a goal but fail to attain it. As we saw in the previous chapter, clause (i) of Wright's account—that S does B for G only if in the actual conditions B tends to bring about G—simply is not true. Just as rational agents may do the wrong thing because they have false beliefs, nonrational agents may act incorrectly because what they do presupposes conditions that do not in fact obtain. As Wooldridge informs us, the wasp goes into the burrow "to see that everything is all right." That is, she checks whether the burrow has been disturbed so as not to lay her eggs in the presence of a predator, and she does this for the sake of assuring healthy and happy young. But in the circumstances of the experimental situation, what she does for this end—check the burrow each time the *cricket* has been moved—does not tend to bring about healthy and happy young. Indeed, the wasp never lays any eggs at all.

Any account of goal-directed behavior must make comprehensible how a particular behavior, such as the wasp's investigation of the burrow, can have a goal that it does not even tend to realize. Central to the judgment that the behavior of the wasp has a goal is the possibility that the behavior may fail to achieve that goal. And central to the kind of normativity embodied in this potential for failure is the possibility that what the wasp does, in the actual circumstances, is counterproductive to reaching the goal—but this counterproductivity in no way cancels out the fact that the behavior has that goal.

Wright's account is incapable of making this type of failure understandable because it applies the criterion of appropriateness with too narrow a scope. Wright's formula T is rooted in the supposition that *each* particular goal-directed behavior must be appropriate to attaining its goal, in the sense that each behavior must tend to bring about its goal. But this condition is too strong. It does not allow for the possibility that a particular behavior can be wrongheaded because it is *inappropriate* to the actual situation but nevertheless can have as its object the very goal that it is inappropriate to accomplish. This is clearly possible in the case of rational behavior that we think of as based on the false beliefs of the agent, as when I attempt to screw an English bayonet lightbulb into its socket. It is also what is going on in the example of the *Sphex* wasp that attempts to build a burrow out of unsuitable material or keeps checking for predators when there are only biologists about. In such instances, the *individual* bit of behavior is inappropriate to bringing about its end, even in the probabilistic sense. But the single bit of behavior still *has* that end in virtue of its relations to other things that the agent does and its role in the *mostly* appropriate behavior of the agent.

Consider, once again, the *Sphex* wasp. What she does when fooled by the biologist clearly fails to attain its end. But for this behavior to be inappropriate, it must be evaluable in terms of a goal, as behavior that has no goal is neither appropriate nor inappropriate. On the other hand, whatever it is for this behavior to have a goal, it can't involve the statistical likelihood that in the actual situation, the behavior will bring about that goal. Wooldridge has stacked the deck. Well, what is the wasp's proximate goal in this situation? The correct nested description of what the wasp is *trying* to do here is surely to check for the presence of predators for the sake of protecting her potential young. But what leads us to believe that the wasp repeatedly goes into the burrow for the sake of checking for predators is not that she always does something in the actual circumstances that tends to result in the attainment of the goal, as Wright would have us believe. The wasp does nothing that has this property. Rather, what convinces us that the wasp is checking for predators is everything else that the wasp does: building the burrow, searching for crickets, and paralyzing them so that she can have healthy and happy young. These behaviors form a complex pattern organized in such a way that, were each act successful, the result would be healthy and happy young. And *most* of what the wasp does *is* appropriate to the ends of those acts in the most straightforward sense possible: those behaviors actually bring about those ends, even though the checking behavior is a failure.

The particular act we are considering, the behavior of going into the burrow, would fit into this pattern if the goal of this particular behavior were to check for the presence of predators so as to avoid laying eggs in their vicinity. Furthermore, there are easily imaginable counterfactual circumstances in which the wasp's viewing the inside of the burrow would be an effective way to avoid laying eggs in the neighborhood of predators. It is these facts that warrant the judgment that the wasp's going into the burrow is appropriate for checking for predators and that the ultimate goal of this activity is healthy young.

In effect, Wright has made a mistake regarding the scope of some quantifiers. For an individual bit of behavior to have a goal, it need not be successful at attaining that goal. Instead, a particular behavior need only be a constituent part of an ongoing series of acts of an agent, most of which are actually successful at attaining their ends. It is not the case that, if the wasp is trying to check for predators, what it does in the situation must succeed in warding off predators. Rather, if the wasp is trying to check for predators, what it does in the situation must fit into a series of other, mostly successful acts by the wasp that, overall, have some other goal. This act must fit into the series in such a way that if it were successful at warding off predators, it would be instrumental in accomplishing the overall goal of the wasp's behavior, and there must be readily conceivable counterfactual circumstances in which the act could be successful at bringing about this end.

This principle, that no act of an agent can have a goal unless most of what that agent does is successful at attaining its goals, is the original principle of charity. It is rooted in the fact that the goal of an individual bit of behavior is a matter of how the behavior is related to the rest of the behavior of the agent and the role of that act in the achievement of the agent's goals, rather than a function of any relation between the behavior and its goal—statistical, causal, or other.

## 2. GOALS AS STRONGLY HOLISTIC PROPERTIES

When one says that some event has a goal, one asserts that some three-place relation obtains among an event, an agent, and a goal. Wright's account attempts to preserve this appearance by suggesting that if some event has a goal, then it (the event) is likely to bring about that (the goal). But such an ontological understanding of the property of an event's having a goal must be wrong. The goal of an event is no individual at all, so the property of an event's having a goal can't involve a relation between that event and its goal. Wright's articulation of what is involved in an event's having a goal must be wrong.

From an ontological point of view, there is a second problem with Wright's account. The role of the agent in the logic of goal attributions remains entirely opaque. The structure of Wright's formula T acknowledges that in many ordinary contexts we assume that, if some event has some goal, it is an act of an agent. Formula T asserts that agent S does B for the sake of G iff . . . , so for Wright the attribution of goal-directedness is logically associated with the attribution to the agent of the event considered as an act. But the content of formula T is entirely silent as to why goal-directed acts are normally attributed to agents.

I previously suggested that the better way to articulate the relational character of goals is to concentrate on the relations among the acts of a single agent rather than the supposed relation between a single act and its goal. Simply because we report the presence of a goal-laden event by specifying its goal, it does not follow that what we thereby report is, ontologically speaking, a relation between that event and its goal.[2] Given the ontological oddity of a relation between an individual and a mere possibility as well as the failure of Wright's formula to capture the role of the relations among an agent's acts in goal ascriptions, the evidence is overwhelming that the goal of an individual bit of behavior is a matter of how that behavior is related to the rest of the behavior of the agent and the role of that act in the achievement of the agent's goals, rather than a function of any relation between the behavior and its goal.

Consider the behavior of the plover when she discontinues the broken wing display. According to Griffin, "[F]inally . . . it flies away rapidly, usually in the same direction, then circl[es] back to the general vicinity, though seldom to the exact spot, where the eggs or young are located." Abstracted from the context that this bit of behavior follows hard on the heels of an episode of predator distraction, such behavior makes little sense, teleological or other. Why should the plover take an indirect route to the spot she comes to? And why should she come to that spot rather than some other? Placed in context, however, the teleological answers to these questions become obvious. The plover initially takes wing in the presence of the intruder, so she flies away from the eggs in order to deceive the intruder regarding her destination and to keep her young concealed. Then she flies back to the vicinity of the young so as to be able to further protect them, but only after she is out of sight of the intruder. Her landing at a spot some distance from the eggs has the goal of avoiding tipping off this intruder, or any other, to the location of the young. The goal of each behavior, taken by itself, is entirely opaque. Taken as an element in temporally extended series of acts, each behavior makes sense in that it provides the conditions requisite for the efficacy of the following act, and all fit together in such a way as to provide for the attainment of the ultimate goal of protecting the young plovers.

Now consider a failed piece of goal-directed plover behavior. On the approach of an intruder, the plover mother displays herself prominently, in a manner that is quite uncharacteristic of plovers that are not mothers. The plover then starts walking away from the nest, but the intruder does not follow (perhaps because the intruder is a field teleologist checking on the goal-directedness of her behavior). The plover stops moving away. She then approaches the intruder, making noises that are characteristic of plovers in distress. In light of this subsequent behavior, the field teleologist concludes that the previous bit of plover behavior, the display and motion away from the nest, represented a failed goal-directed attempt to induce the intruder to follow the plover for the sake of diverting the intruder from her eggs. The teleologist bases this judgment on the fact that when the plover's initial routine is interrupted, she does something else that would make sense if it had the goal of diverting the intruder from the eggs, even though in the actual circumstances (the presence of a stubborn teleologist) it has no such effect. This evidence clearly supports a counterfactual element, just as Wright suggests. But he has misidentified this element. According to Wright, this sort of evidence supports the counterfactual judgment that, had initial circumstances been different, the agent would have done something else which, in those circumstances, would have tended to bring about the state of affairs that is the goal. But, given the actual circumstances, the plover's behavior after the initial failure does not in fact tend to bring about the end of leading the intruder away. Instead, the plover does something else that teleologically makes sense as having the goal of leading the intruder away. To say that a behavior "makes sense" in context if it has a certain goal is to say that two things are true. First, the goal assigned to both of the failed acts is instrumental to what appears to be the ultimate point of the plover's previous and subsequent behavior, protection of her nest. And second, what the plover in fact does *would* be successful in bringing about this instrumentally valuable result in many circumstances, but not in the actual circumstances.

The attribution of goals to bits of plover behavior thus traces a distinctive pattern. Each bit of behavior is assigned a goal in light of the goals assigned to other behavior, both short- and long-term. The most obvious constraint on these collective attributions is that most of the goals assigned must be goals that the behaviors actually achieve. If this constraint is not satisfied, the entire pattern will collapse. To paraphrase a remark that Davidson has made in a different but analogous context, the possibility of failure (error) makes sense only against a background of massive success (truth).

This pattern of attribution of properties to separate, though related, individuals is typical of a particular sort of relation. As Jerry Fodor and Ernest Lepore, among many others, have pointed out, some properties are such that it is possible for only a single thing to have them.[3] Being red and being a square are examples. Although many things are in fact squares, there is nothing about what it is to be a square that precludes the possibility that only one thing can have this property. Being a sibling, however, is different. Because what it is to be a sibling involves a reciprocal relation between two different entities, if one thing has this property, it follows that at least one other thing does as well. Similarly, but for different reasons, there cannot be only one entity that is a natural number. To be a natural number is to have a successor that is itself a natural number but is not identical with the initial entity. So if one thing is a natural number, others must be so, too. Indeed, given what it is that qualifies an entity to be such a number, if there is one natural number, there are endlessly many. If a property is similar to the property of being a natural number in the sense that if there is one thing that has that property there must be very many such things, we say that the property is holistic.

The property of being a natural number demonstrates in a clear fashion the feature of holistic properties that guarantees that they are holistic. All holistic properties are relational. But only some relations are holistic. To be a father is relational. Thus, if one thing is a father, there must be at least one other thing in the world, a child. But there is nothing about being a father that guarantees that the child must also be a father. To be a number is also to have a certain relation with something else. But the character of that relation (having a successor that is another natural number) guarantees that that other thing must also display that same relation, in this case, to some third entity.

Natural numbers also display another feature that some, but not all, holistic properties possess. In the case of natural numbers, *which* number a particular entity is, is determined by which number it is related to. It is not merely the case that 4, for example, could not be a number unless it had the successor it does. Unless its successor was 5, it could not be 4. We will call properties that display this feature "strongly holistic." And, given that goals are attributed in this way, we can conclude that having a goal is a strongly holistic property.

For a property of an individual to be strongly holistic is for it to be such that the property is holistic and the specific character of that property is fixed by corresponding holistic properties in the individuals to which this one is related. Because the goals of other events are relevant to the attribution of a specific goal to an individual event, we are warranted in attributing that goal only if we also have reason to believe that those other events have the goals they do. The argument through which we assign a definite goal to an event always relies on premises relating to the goals of other events. So the evidence in favor of an event's having a goal is strongly holistic. That is, one can have reason to believe that some event has a specific goal only if one also has reason to believe that other events have the specific goals that they have.[4]

It might seem that identifying goal-directedness as a strongly holistic property (or at least a property for which the evidence is strongly holistic) has only made the ontological and epistemological problems involved in goal attributions more serious. To determine whether some particular event has a goal, one must consult the evidence to see whether it is part of a pattern of goal-directed events. But if one can't tell whether a single event has a goal without knowing whether multiple other events have goals (and fit together into a pattern), how could one ever know that any individual event has a goal? For mustn't one be able to determine that one event has a goal before one can make the judgment that more than one event has a goal? Furthermore, speaking ontologically rather than epistemologically, how can it be the case that no event can have a goal unless many do?

This problem is familiar from other contexts in which philosophers have attempted to make sense of strongly holistic properties. In one version, it is the problem of the hermeneutic circle. The traditional form of the problem assumes that meaning is a strongly holistic property, one in which the meaning of some utterance U depends on the meaning of a large number of other utterances. The problem becomes obvious in both its epistemological and its ontological variations. How can one know the meaning of any given utterance U, if in order to know it one needs to know the meaning of a large number of other utterances and if one can't know the meaning of any one of them unless one already knows the meaning of yet others? Similarly, what does it mean to say that no single utterance can have a meaning unless a large number of utterances do? Don't a large number of utterances have a meaning only because individual utterances do?

The last, ontological version of the hermeneutic puzzle points toward a solution, both in the standard case of meaning and in the case of goal-directedness. One must remember that one is dealing with holistic properties and that holistic properties depend on a set of relations among the entities that have those properties. One can't be a participant in a circle dance unless there are other participants in that circle dance. There is nothing mysterious about this statement. To be a participant in a circle dance is to have a certain relation to some other person or persons, a relation that guarantees that if you are a participant, the others to whom you are related are participants as well. So, ontologically speaking, each has the property only if many do. Whatever justifies the judgment that one person is a participant must also justify the judgment that some others are as well. To have the property is to be in a certain relation to others, and to be in that relation implies that those others have the same property. Whatever counts in favor of one's having the property also counts in favor of others' having it. Thus the knowledge that multiple entities have the property in question *can't* be derived from the knowledge that single individuals have that property.

Logically speaking, there is nothing odd or aberrant about the thought that an event's having a goal is a strongly holistic property that depends on relations among a large number of events that also have goals. But this statement leaves open the character of the relations among events that constitute goal-directedness.

## 3. LIVING AGENTS AND TELEOLOGY

What is the role of the agent in our teleological discourse? In most ordinary contexts, we assign a goal to an event only if we understand that event to be the act of some agent. Wright's formula T assumes this fact in its very structure, as it attempts to give necessary and sufficient conditions for "S does B for G." But nothing that Wright says gives us any insight into why goals and agents go together.

We can work our way toward an understanding of the centrality of an agent to goal-directedness if we go back to the fact that motivated the judgment that goal-directedness is a strongly holistic property of events. Some acts have a goal that they don't attain. This can happen, for example, because an act presupposes a counterfactual state of affairs, so that what is done does not even tend, in the actual circumstances, to bring about its goal. In the last section, I argued that this failure forces us to conclude that what it is for an individual act to have a goal must involve holistic relations among a set of acts. It is only if what it is for a given act to have a goal is for that act to be appropriately related to other goal-directed acts that we can understand how an act can be inappropriate to bring about an end and yet have bringing about that end as a goal.

That the acts of agents are frequently misguided given the actual environment is thus of crucial importance to the understanding of goals. This potential for failure implies that there is a difference in principle between whatever it is for some event to have a goal and the actual effects of that event. That an event has effects E, or even tends to have those effects, can't be a necessary condition on that event's having E as its goal. It is this gap in principle between the goal of an event and what that event in fact brings about that allows for the normative evaluability of goal-directed acts. For what an act *should* bring about is its goal, even though there is no guarantee that it *will* bring this about.

This normative character of goals provides the best opportunity for understanding the role of agents in goal-directedness. A difference in principle between the goal of the event and its actual effects is a necessary condition on any event's having a goal, even though goal and effect often coincide. This difference in principle is central to the normative character of goal talk. And, I will argue, it is a necessary condition on this difference in principle that events with goals are acts of a certain kind of agent, a living one.

A norm is a standard, model, or pattern that is used in an evaluation. This evaluation doesn't just measure whether some individual is similar to the norm. Dissimilarity to the norm implies a falling short, a failure, whereas similarity implies success. Thus, words such as "bad," "good," "ought," and "should" are applied in light of evaluations according to norms. Furthermore, norms involve conformity to type. A norm is applicable in an evaluation only if the individual evaluated belongs to the type governed by that norm. So, if something is a bad apple, it is an *apple* that, when compared with a norm for being an apple, falls short in that it ought to be more like the norm.

In most cases, the norm has a source external to the individual evaluated. If an apple is a bad apple, for example, it might be so relative to a norm for apples that is established in light of the potential consequences for human beings eating an apple: it ought to be such that, when eaten, it nourishes and delights human beings. But there is nothing intrinsic to the apple itself, or to apples as such, that determines that this is to be the norm according to which apples ought to be evaluated. In fact, most uses of evaluative terms are relative. To use Heraclitus's example, salt water is both good and bad. It is good for fish and bad for humans. There is no contradiction here, because the norms used in the evaluations are rooted in different possible effects of ingesting the same water for humans and for (saltwater) fish. There is nothing intrinsic to the salt water that determines, on a higher level, which norm ought to be applied. That is, there is usually no norm rooted in the entity evaluated that decides among potential norms for evaluating that entity. In this sense, the norms used in most evaluations—and the evaluations themselves—are arbitrary.

Virtually any individual can be evaluated normatively, and this is as true of events as of anything else. Events involving human beings are often evaluated instrumentally. In such cases, the norm derives from whether the effects of that event bring about some further result. If I fall off a horse and get a concussion, this is bad in the sense that the event is bad for me. The end, or result, in question is the continued good functioning of my organism, and the norm for evaluation rooted in this end is something like "events that do not have detrimental effects on the functioning of Okrent's body." We can apply the same sort of evaluation to acts. If I want to get to Lewiston from Portland, then I should take Route 26. That is, the act of driving on Route 26 is instrumental to bringing about my being in Lewiston, and, relative to this possible state of affairs, driving on Route 26 is the right thing to do. On the other hand, if you for some reason want me to be in Biddeford, then, relative to that end, I ought not to take Route 26. As far as this story goes, there is no reason to think that either the end of being in Lewiston or the end of being in Biddeford is grounded in anything about the actual driving itself. In such cases, the norm for evaluating the event is rooted in the desires of the evaluator, and there is nothing that determines which of those norms, or which of those desires, ought to be regulative.

If we change our perspective on these very same events, however, we see a very different situation. My driving on Route 26 is either instrumentally good or instrumentally bad, depending on the results of that event and the result that is desired. There is no basis on which to choose between my desires and yours as grounds for the establishment of a norm for evaluation of the instrumental results of the event. But we can also approach this event from the standpoint of a request for an explanation why the event occurred. From this perspective, my desire does have a priority. That I desired to go to Lewiston helps to explain why I drove on Route 26, whereas your desire for me to be in Biddeford doesn't (or at least it doesn't do so directly). That is, in the case of rational explanation, it is only the desire of the agent that is relevant. The event happens *because of* that desire. Furthermore, in the case of rational action, the desire indexes the *goal* of the event: in general, if B is an act of S that is explained by S's desire that G, then G is the goal of B.

The desire to get to Lewiston doesn't explain my driving on Route 26 all by itself, of course. I can have this very same desire and not drive on Route 26 at all, if, for example, I believe that driving on Route 1 gets me there. It is the combination of desire and belief that rationally explains the act. It does so by fixing both the goal of the act and the conditions under which the act would succeed (the content of the belief). Why did the driving on Route 26 occur? Because it would get me to Lewiston if it were an open road between Portland and Lewiston. The event of driving on Route 26 is in order to get to Lewiston, and it happens because it is for getting to Lewiston (even though, of course, it might fail to get me there).

The event of driving on Route 26 can be typed in any number of ways. But, if it is explained rationally, it occurs because it has the goal of getting me to Lewiston. So there is a perfectly legitimate sense in which this type, being for getting me to Lewiston, fixes the identity of this event. If what a thing is, is fixed by the feature of that thing that explains why it occurs, then, if an event happens because it has the goal of bringing about some result, what that event *is,* is an event for bringing about that result.

If this is the case, then there is a nonarbitrary norm for evaluating rational actions. If driving on Route 26 is essentially for getting me to Lewiston, then the norm for evaluating that event is *getting me to Lewiston.* That is, for a rational act, there is a sense in which the goal of the act fixes the identity of the act, and that identity in turn fixes how and against what norm that act is to be evaluated. If my act of driving gets me to Lewiston, it is a success; if not, it is a failure.

This fact is reflected in our ordinary language. Whenever we speak about the goal of a rational act, it is perfectly clear that, regardless of how else that act might be evaluated, there is a sense in which the goal itself fixes the standard against which the act is to be judged. Even if we would prefer that the agent not have the beliefs and desires that motivate the act, or even if we think that the actual consequences of the act are bad (from our own or from a moral perspective), or even if, from the long-term perspective of the agent, the act is bad in that it is not in her interests or has an end that thwarts her deepest desires, we fully understand the usage that evaluates the act from the standpoint of the act itself.

Events have goals only if there is some goal of the event that fixes a nonarbitrary norm against which the actual effects of the event can be measured. Such a nonarbitrary norm must be rooted in the identity of the event, and that identity is fixed by the fact that the event happens because it is the sort of event that it is. In the case of rational action, the goal of the act, which fixes both the identity of the event and the norm against which it is to be evaluated, is indexed by the belief and desire of the rational agent: the event happens because it is for satisfying the desire of the agent.

In the case of nonrational behavior with a goal, if there is such a thing, this indexing is impossible. If there is a legitimate nonmetaphorical sense in which the wasp does things with the goal of reproducing, then what it does has a goal independent of the wasp's having any beliefs and desires. This is so because we are confident that the wasp is nonrational and does not act for reasons. Nevertheless, if what the wasp does has a goal, this goal must fix a nonarbitrary norm against which what the wasp does can be evaluated; this norm must be rooted in the identity of the wasp's acts; and this identity, in turn, must arise out of the fact that what the wasp does happens because it is for the sake of realizing its goal. All of this is in accordance with ordinary usage. Even though there are many ways of describing the current behavior of the wasp, we say that what it is doing *is* building a burrow and that this action happens *because* it is a building of a burrow.

This description leaves us with a problem. In the absence of beliefs and desires, what can fix the identity of an act in such a way as to explain it? On the surface, the most direct approach to this question is by way of the character of teleological explanation. As it turns out, however, the best way to understand goal-directed explanations is by first comprehending how the identities of the acts of agents are fixed.

That an event is the act of some agent is part of what it is for an event to have a goal. This much is common to ordinary usage and to Wright's formula T. By why should this be so? Consider, yet again, the *Sphex* wasp. Part of the context in which we fix the identity of the event that we describe as the wasp's going into the burrow "to see that everything is all right" is what else the wasp has been up to. But this surely isn't the entire story. Taken together, the various behaviors that supply the context for the interpretation of the checking behavior make sense as the acts of a *wasp*. We also take it to be characteristic of wasps that they can be expected to act for the sake of attaining goals of their own. Among these goals are survival and *reproduction*. Many of the things that wasps do, including the behavior under description here, make sense in terms of the overall goal of reproducing. It is only in light of this estimation of the goals of the agent that we offer our interpretation of the proximate goals of its acts.

In attributing a goal to an act of an agent, we appeal not only to the goals of other acts of the same agent but also to the goal *of that agent itself.* The reasoning seems to be as follows. This thing is a wasp. A wasp is a being of a sort that has the goal of reproducing, although not all wasps act with the goal of reproducing and even those that do so act with the goal of reproducing do so only periodically. When acting on this goal, this agent will do things in order to reproduce. There are several sorts of circumstances that are necessary for wasp reproduction. For example, since wasps reproduce by laying eggs that hatch into grubs that cannot move, the young need a ready source of food. Since the young are incapable of predator avoidance, they must be well hidden from potential predators if they are to survive. Much of what the actual wasp actually does will bring about actual circumstances of the sorts that are necessary for wasp reproduction. This fact, combined with the fact that the agent is a wasp and the fact that what it is to be a wasp involves acting so as to reproduce, provides the context in which it is reasonable to conclude that the goal of the failed wasp behavior was to check for predators.

Living agents have other typical goals. *All* living agents can be expected to act for the sake of staying alive, for example. We ordinarily take for granted that this kind of goal-directed behavior usually takes two forms, an external form and an internal form. An animal, for example, can usually be expected to interact with its environments in such a way as to keep itself alive, and this fact is reflected in the nested teleological descriptions we give of its overt activity. Cold-blooded reptiles that need to keep their body temperature within certain limits in order to survive are said to seek shade on excessively hot days, so as to cool their body temperature for the sake of staying alive. Herbivores that need to eat plant food in order to survive are said to search for suitable plants so as to eat for the sake of staying alive. And carnivores that need to eat meat in order to survive are said to hunt so as to kill other animals so as to eat for the sake of staying alive. Similarly, events that occur in the internal organs, or functional parts, of organisms—animals, plants, microbes, and fungi alike—are given nested goal-directed descriptions culminating in their doing what they do to fulfill some requirement of the life processes of the organism. Hearts engage in behavior that is described as for the sake of pumping blood, which in turn has the goal of oxygenating tissues, which itself is necessary for survival.

The attribution of a goal to a particular bit of behavior thus involves contextualizing that behavior in two related ways. First, on the basis of morphology and overall behavior, we identify the agent whose behavior we want to understand as being of a certain type, where instances of that type have characteristic goals and characteristic ways of attempting to achieve those goals. This agent is a wasp; wasps attempt to reproduce by laying eggs in safe and food-rich environments. Second, we examine the actual behavior of the agent to determine the extent to which the agent's actions can be interpreted as an attempt to achieve these characteristic goals in this characteristic way. This is possible only to the extent that most of the agent's acts can be seen as successful at achieving its goals, if they are interpreted as having the goals that they would have were they all, taken together, interpreted as an attempt to achieve one of the agent's characteristic goals in a characteristic way. We interpret this behavior in the way we do because most of what the wasp does is so very good at achieving results that, taken together, would achieve the goal of wasp reproduction that we take to be typical of wasps. We then proceed to interpret the behavior of the wasp that does not fit into this pattern of success—but would fit, were conditions slightly different—as having the same goal that it would have if it did fit the pattern. The goal of going into the burrow is to check for predators so as to ensure safe eggs, even though under the circumstances the wasp's actions do not realize this end, because, in the context supplied by the rest of the wasp's behavior, this behavior would facilitate wasp reproduction by achieving this goal, were conditions slightly different.

Thus, the pattern of relations in virtue of which an individual act is said to have a goal is a pattern of relations among the acts of an agent *and* between those acts taken together and the goals of the agent itself. It is only through an appeal to the intrinsic goals of the agent that the underdetermination of interpretation inherent in the ascription of strongly holistic properties can be overcome. In the absence of such a reference to the goals of the agent, any attribution of a goal to a particular piece of behavior will be relative to the interests of the attributor and, in that sense, arbitrary. This is precisely the case when we interpret the behavior of machines in a goal-directed fashion. The difference between a device that is bad at fulfilling the goals of computing and good at satisfying the ends of anchors is nothing more than the interests of the one doing the evaluating, for the device itself has no goals that can fix the ultimate end in terms of which its behavior is to be understood.

But by what right do we conclude that an agent has a goal of its own? We do this in one way, of course, when we appeal to the desires of rational agents. If an agent is acting rationally so as to achieve a goal, then that agent desires that goal, and, in that sense, the goal is a goal of the agent rather than a goal that is merely arbitrarily used to evaluate the agent's behavior. But can nonrational agents have ends of their own in the absence of mental states with content? Or are all such attributions mere metaphoric projections from the case of rational agents?

An individual event has a goal only if there is a nonarbitrary norm against which that event is to be evaluated. Such a nonarbitrary norm exists only if the event is an act of an agent that has some goal of its own, a goal that is being facilitated by the act. This goal supplies a norm against which the overall behavior of the agent can be evaluated. If the second norm is arbitrary, then so is the first. As we saw above, the nonarbitrary normative evaluability of any individual, whether an act or an agent, depends on whether the very being of that individual prescribes the norm against which it is to be evaluated. But for an individual to have a goal at all, there must be such a nonarbitrary norm. So to determine whether any nonrational acts have goals, we must first determine whether there are any nonrational agents whose very being prescribes a nonarbitrary goal, or norm, for that individual.

Are there any such agents, evaluable in themselves? Prior to the seventeenth century, there was a standard answer to this question, rooted in the Semitic religious tradition. Every finite thing that is, is a creature, created by a maker. That maker made each thing according to a plan. This plan constitutes the essence of that thing and is the standard according to which the creature is to be normatively evaluated. After the scientific revolution of the seventeenth century, this way of thinking fell from favor outside of biological contexts. One can explain the existence and behavior of physical types of things without appeal to a maker. After the publication of Charles Darwin's *On the Origin of Species*, this way of thinking fell out of favor even in biology. Darwin shows how biological entities could be as they are even in the absence of a designer. Does the demise of creation as a scientific concept signal the end of the goal-directedness in nature?

In biology, at least, it does not. What, after all, did Darwin explain by appeal to natural selection? Reproduction under selection pressure explains why there are *successful* individuals in the world; objects that are well designed, organs that function admirably, acts that achieve their goals. As Daniel Dennett notes, "The key to understanding Darwin's contribution is *granting* the premise of the Argument from Design."[5] If natural selection explains anything, it explains the prevalence in the world of individuals that are *appropriate*. But if appropriateness is judged against an arbitrary standard, there is nothing to explain. Every individual both satisfies and does not satisfy an infinite number of such standards. So the explanandum of evolutionary explanations presupposes that there are at least some individuals that are evaluable, in themselves, according to some nonarbitrary criterion.

But which individuals are these? And in virtue of what characteristic of these individuals do these norms apply to them? Darwin is often described as anti-Aristotelian. And, indeed, there are many respects in which he undercut Aristotle's legacy. Nevertheless, there is a central respect in which Darwin was the greatest Aristotelian of the nineteenth century. Darwin agrees with Aristotle—and disagrees with Christianity—on the central issue of whether individuals are evaluable in a nonarbitrary fashion even if they were not made by some rational creator. Darwin even agrees with Aristotle in his judgment concerning which things are so evaluable: living things. For Darwin and Darwinians, *living* organisms are those individuals that carry the principle of nonarbitrary normative evaluability in themselves. Nonarbitrary standards for evaluating goal-directed events are borrowed from nonarbitrary standards for evaluating the entities in which they occur.

At first sight, there is something surprising about this claim. There is a long tradition, after all, that finds it more intelligible, for example, that an organ should have a function than that the animal or plant that has the organ should have a function. The function of eyes is to see, which is good for the animal. So particular eyes can be evaluated according to how well they perform the function of seeing. But what is the function of the animal, according to which it can be evaluated? That this should be problematic is itself a function of the fact that the criteria for judging the parts of organic entities, whether those parts are spatial (organs) or temporal (processes and acts), must be derived from the criteria built into what it is to be the organism of which they are parts, but the standards for normatively evaluating those organisms cannot be understood in the same way.

But we are not interested in the *function* of an organism. We are interested in its *goal:* what it is working to achieve. For it is this goal that grounds attributions of goals to particular acts of the organism, and it is much more natural to think of a living organism as having a goal than it is to think of it as having a function.

The traditional concept of a living organism assigns to living things the fundamental job of staying alive. That a living organism acts in order to stay alive has traditionally been understood to be part of what it is to be alive at all, a job that is built into the fact that it is alive. To be alive is to work at being alive. Consider, for example, this entirely typical passage from Richard Dawkins:

> The minimum requirement for us to recognize an object as an animal or plant is that it should succeed in making a living of some sort . . . . It is true that there are quite a number of ways of making a living—flying, swimming, swinging through the trees, and so on. But, however many ways there may be of being alive, it is certain that there are vastly more ways of being dead, or rather not alive. You may throw cells together at random, over and over again for a billion years, and not once will you get a conglomeration that flies or swims or burrows or runs, or does anything, even badly, that could remotely be construed as working to keep itself alive. . . . Staving off death is a thing you have to work at."[6]

The view expressed in this passage is that an object is alive just in case it does something that succeeds in keeping it alive, that is, something that continues to succeed in keeping it alive, and so on.

Given this notion of life, for a thing to be alive is for it, in itself, to be subject to a specific norm. A living thing succeeds just in case it continues to go on living. Otherwise it fails. Since to be alive is to do something that allows it to continue to live, it is life itself that provides the standard against which a particular living thing is to be evaluated. The end or goal of the process is the very process itself, for the end of the process is that the process should continue.

This sense that the standard by which life is to be evaluated is the continuation of life itself and that life itself is to be defined in terms of a continuing process, goes very far back in the tradition in which Dawkins writes. Compare this passage by Aristotle, in which he distinguishes mere motions from "actualities" (*energeia*), actions in which the "end is present in the action," such as life. "At the same time we are living well and have lived well, and are happy and have been happy. If not, the process would have had sometime to cease, as the process of making thin ceases; but as things are it does not cease; we are living and have lived."[7] Building (another of Aristotle's examples) and making thin are processes in which the end is not in the process itself. When the end is attained, the process ceases. Living is not like that. The end of living is attained only insofar as the process with that goal continues, for the end of the process is the process itself.

There is an air of vacuity, infinite regress, or circularity about such an account of life. To say that life is a process in an object that succeeds in continuing the very same process is to invite the response that, in this view, any old repetitive process marks the entity in which it occurs as alive. The fact that the motions of the planets in the solar system continue until they set up conditions that result in a repetition of those same types of motions, does not mark the solar system as a living thing; but this account of life suggests that we should consider it to be alive. So being an entity that undergoes or undertakes a process that continues itself may be a necessary condition on being alive, but it cannot be a sufficient condition.

The traditional concept of an organism can be supplemented with the seminal work of the physicist Erwin Schrödinger.[8] His basic intuition was that living things were entities that maintained an ordered structure in a variety of situations by doing something that amounted to engaging in interactions with its environment. To say that living things have an order is to say that such things are in disequilibrium with their environment, that they have physical properties markedly different from those of their surroundings. The second law of thermodynamics suggests that such disequilibrium will tend to disappear and that entropy will tend to increase. This law holds for the universe as a whole. But in certain special conditions, there can be local exceptions to the law, though such exceptions are inherently unlikely. When such disequilibrium occurs and is maintained over time, one has a local exception to the law and a perpetuation of order. Such a perpetuation of order can be maintained, however, only if there is a continual interchange between the entity that displays that order, on the one hand, and its surroundings, on the other, an interchange that both replaces the building blocks of the physical parts of the ordered entity and maintains the ordered structure of the entity that undergoes these changes.

Schrödinger suggests that any entity that satisfies these conditions counts as alive. It is alive only insofar as it keeps working at being alive, that is, continues to display those changes through which its order is maintained. And it continues to work at being alive only insofar as it maintains the structure of disequilibrium with its environment that causes it to undergo these changes. As Dawkins would say, the living thing is the thing that works at being alive. As Aristotle would say, living is an act whose end is in the act. But life can be further characterized. A living thing has an ordered structure that is maintained in a variety of environments by undergoing specific interactions with those environments, even though it is inherently unlikely that the structure will be maintained, given only the global direction of the laws governing the universe.

In what sense, if any, does an entity that is alive stand under a nonarbitrary norm? In what sense, if any, does it have a goal in itself? To describe an entity as alive is to attribute to it both a *structure* that has been and is maintained and a *pattern* of events in the entity, a metabolism, or process of interchange with its environment, through which the structure is maintained. It is thus part of the essence of the living thing as living that it undergo a certain series of changes. The most general description of this series of events is that, insofar as a thing continues to live, *something* must occur such that, in the actual conditions, that series will repeat and the ordered structure that is associated with the repeating pattern of events is maintained.

*If* this occurs, the living thing remains what it is, a living thing of a sort specified by its structure and metabolism. To be alive is to do something such that these patterns continue. But, even if it is necessary that the living thing undergo such a series of events for it to stay alive, this does not imply that it is necessary that it in fact undergo such changes. At any moment, the life process *might* cease. So to describe a thing as living is to specify what has been the case and what would be the case were the living thing to continue to be itself, to be alive. But it is not to specify what *will* be the case, although of course the pattern might persist. When one considers life from this future-oriented perspective, the pattern of the entity's life functions emerges as a standard or norm—though not necessarily a realized one—against which what actually does happen can be measured.

This norm is also a goal. Insofar as the thing is alive, what it does admits of the description that it is something that, in the actual conditions, results in continued life. But, for all that, there is no necessity that what is done, in actual conditions, will result in attaining this result. A living organism will act *for* life; it will do something that would result in further life, were the conditions right. An entity that is alive in this sense has a goal in itself, an inherent standard that can be used in evaluating it. It is successful insofar as it continues to engage in those processes that both characterize it and maintain its structure, that is, insofar as it continues to live. When the activity ceases, and the order no longer is maintained, it fails; it is dead. Any entity that displays these features, regardless of its chemical and physical constitution, counts as alive, in this sense.

To describe a thing as alive is to suggest a perspective toward it that privileges a certain pattern of events in that entity and a certain structure that has been maintained in that entity. That pattern and that structure are privileged insofar as they are *considered* as norm and goal against which are judged what happens to that entity and what it does in the future. There are, of course, other ways to describe any living thing. One can even use "life" as an entirely descriptive term for specifying what has happened in the past, as opposed to treating it as a specification of what *should* continue to happen in the future. So having a nonarbitrary goal can't be reduced to being alive. But, nevertheless, there is a clear correlation between the occasions when we are willing to ascribe a nonarbitrary goal to an entity and those when it is correct to say that the thing is alive.[9]

Agents that are alive can have goals other than the goal of staying alive. In particular, living things can have the goal of producing other individuals of the same type as themselves, that is, the goal of reproducing. As with the goal of staying alive, this goal of the agent must be seen as intrinsic to the agent's being the sort of agent it is. As with the goal of staying alive, the goal of reproduction serves as a nonarbitrary anchor for attributions of goals to specific behavioral patterns of the organisms that act for this goal. As reproduction raises important issues regarding explanation, I will discuss this goal more fully in the next chapter, which concerns the logic of goal-directed explanations.

We can now see why all goal-directed behavior is the behavior of an agent that itself has a goal. An event has a goal only if it stands under a nonarbitrary norm, a norm that establishes the goal of the act. Such nonarbitrary norms for the evaluation of acts depend on relations among the act, other acts of the same agent, and, crucially, the goals of the agent itself. *Agents* have goals only if what they are, their being, specifies a norm against which the agent is to be evaluated. And only living things stand under such nonarbitrary norms and so have goals of their own. It thus follows that only the acts of living agents can have a goal.

As Dawkins suggests, there are many different ways of making a living. That is, there are many different ways, both behavioral and metabolic, in which different types of organisms achieve the goal of staying alive. The life histories that are typical of different kinds of organisms amount to so many different normative patterns that allow us both to attribute goals to the failed acts of organisms of a given type and to evaluate those acts for success and failure.

The recognition of such normative patterns also allows us to offer a preliminary formulation of the descriptive content of judgments of goal attributions.

S does B for the sake of G only if

> (1) S is a living agent;
>
> (2) S has some intrinsic goal G* such that S's doing something at the time it does B that results in G would, given everything else that S's acts result in, tend to realize G*; and
>
> (3) S is an instance of a type of living agent that typically acts so as to realize G* by doing something similar to B that results in G.

This articulation is meant to do roughly the same job as clause (i) of Wright's formula T, that is, to specify the descriptive content of judgments that assert that some nonrational agent did something that had some goal. This new formulation is much more complex than Wright's attempt. The complexity of the new formulation reflects its attempt to capture the way in which having a goal is a strongly holistic property of an act of an agent that has a goal of its own and is an instance of a type of agent that typically accomplishes that goal by engaging in a certain sequence of acts.

This preliminary formulation articulates what is involved in an act's having a goal through an explicit appeal to another attribution of a goal, the intrinsic goal of an agent. This articulation is not circular, however. What it is for an act to have a goal can be understood only in terms of an agent's having an intrinsic goal of its own, but what it is for an agent to have a goal of its own can be understood independently from any reference to its acts' having a goal. An entity has intrinsic goals of its own if, and only if, it is *alive,* and the conditions under which an entity counts as living can be articulated without any reference to the goals of the acts of the entity. And, if any entity counts as living, that very fact not only ensures that it has some goals of its own but also determines, at least in some cases, the contents of those goals. As Dawkins says, all living things must work at making a living; that is, all living things have the intrinsic goal of surviving. The fact that some S fulfills condition (1) in the above formula guarantees that there is some G*, such that it is an intrinsic goal of S to achieve G*. That S has a *way* of making a living—that it is an organism of some type—guarantees that it has a characteristic structure and metabolism and that, insofar as it continues to live, it undergoes a pattern of changes that both serves to keep it alive and constitutes that life. This in turn guarantees that if some S is a living agent, there is some pattern of actions that it typically uses to attain its intrinsic goal G*, so that any living thing will also satisfy clause (3) in the above formula.

Given the above, all that is required by the above formulation to make sense of the descriptive content of the notion that some S does some B for the sake of some G is an understanding of what it is for S to be alive and an understanding of what sort of living thing S is. But, of course, that some B satisfies the condition formulated above doesn't guarantee that it has goal G. For that to be the case, another condition must be met. As we saw above, goals of individuals are nonarbitrary only if they are rooted in the identities of those individuals, and, in the case of events, identity is established by that feature of the event that explains why it occurs. So an event that satisfies the above formulation counts as having goal G only if the event happens *because* it satisfies that very complex formula. To say this is to make explicit the point that goal-directed descriptions are inherently teleological explanations. But in what sense can an event having this complex, strongly holistic property explain that it occurs? I turn to this question in the next chapter.

CHAPTER FOUR

# THE EXPLANATORY POWER OF GOAL ATTRIBUTIONS

## 1. HOW GOAL-DIRECTED ATTRIBUTIONS EXPLAIN

"To note that the thing that I call my individuality is only a pattern or dance, that is what it means when one discovers how long it takes for the atoms of the brain to be replaced by other atoms. The atoms come into my brain, dance a dance, and then go out—there are always new atoms, but always doing the same dance, remembering what the dance was yesterday."[1]

This passage from the physicist Richard Feynman, in which he comments on the significance of the fact that organisms maintain themselves through exchanges of material with their environments, expresses more eloquently and, in its own way, more precisely than I ever could the view of life that I laid out in the last chapter. Living things do more than dance a distinctive repetitive dance. They *are* distinctive repetitive dances. That a wasp is composed of a collection of atoms of certain types in certain proportions arranged in a certain way is necessary for there to be a wasp, but not sufficient. There is a wasp only if these atoms, taken together, dance the dance of a wasp. This dance involves a sequence of interactions among the various functioning parts of the wasp and between the wasp and its environment. The wasp exists only as long as its characteristic dance continues.

Every living thing dances some such dance, and every living thing is typed according to its dance. Types of dances of life differ from one another in a variety of dimensions; as Richard Dawkins says, there are all sorts of ways of making a living. Different living things have different structures and "metabolisms" that are suited to these ways of making a living. In general, to continue their dances (that is, survive), animals must interact with their environments so as to destroy other living things and to reconstitute their components as functioning parts of themselves, while plants can interact with the nonliving environment so as to produce their own proximate material components. So plants must have structures that transform nonliving components into living ones and, in fact, repeatedly perform the acts through which this transformation occurs, while animals must have structures that transform one type of organic material into another and repeatedly undergo the changes that result in such a transformation. Because animals need specialized, living resources that are not evenly distributed in their environments, it is, in general, helpful to them if their dances include motion relative to those environments. Mere mobility, however, is of no value if the animals cannot distinguish among and react differentially toward those items in their environments that are necessary for the continuation of their lives and those that are potentially damaging. So, in general, it is helpful to animals to have sensory organs that discern differences in their environments. But many different mechanisms can generate motion or discern environmental differences, and each of these mechanisms has corresponding maintenance requirements, so that the mechanism can help to provide food so that the mechanism can be maintained. The dance that the living thing dances thus involves not merely a characteristic sequence of interchanges with its environment but also a characteristic morphology and metabolism that sustain the dance and form constitutive aspects of the dance itself.

The various ways of making a living that are characteristic of living things are more than dances that living things do. This metaphor suggests that the living thing is one thing, and its dance is another. But all these ways of making a living are also constitutive of the livings that they help to make. The blood circulating in the plover is not just instrumental to keeping the plover alive; it is part of what it is for the plover to be alive. The developmental stages that wasp grubs pass through are not just instrumental to the existence of the wasp, taken as a whole; they are temporal parts of what it is to be the wasp, taken as a whole.

These myriad ways of making a living, with their corresponding structures and metabolisms, fall into what I will call "types." Similarities among the ways of making a living occur on as many dimensions as differences among the ways of making a living. Organisms that are superficially similar in their modes of locomotion and sensation can be profoundly dissimilar in their metabolisms and the mechanisms that support those modes of locomotion and sensation. Witness the similarities and differences among birds and bees and bats. Sometimes, however, the similarities among the dances, the modes of life, of a group of organisms—and thus the similarities among their corresponding morphologies and metabolisms—are so pervasive and deep that we classify these organisms as belonging to the same type. As these similarities can be more or less exact and more or less pervasive, these types can be more or less expansive or narrow. Organisms belong to species, genera, classes, orders, and kingdoms, and corresponding to each of these kinds of kinds are different ranges and degrees of similarity and difference.

The terms used to designate individual living things most often reflect the organic types that they embody. When I want to refer to that thing over there, I talk about "the wasp," and when I want to refer to this apparently lame-winged thing over here, I talk about "the plover." Such attributions depend on the fact that the individual that we confront realizes the type that is implicit in the kind terms "wasp" and "plover." Even if Hilary Putnam and Saul Kripke are right, and ultimately the meanings of natural kind terms are not to be solely cashed out in terms of a description, the types that these terms pick out are styles of life. A plover is a plover because it realizes a plover kind of life by having a plover sort of morphology and metabolism, and a wasp is a wasp because it realizes a wasp sort of life by having a wasp kind of morphology and metabolism. So every correct attribution of a type to an organic individual and every appropriate reference to an individual using an organic type term contains an implicit characterization of the organic style that the life of this individual realizes. We can be wrong regarding whether something is a wasp; and we can be wrong regarding the style of life that is characteristic of wasps; but if the individual is a wasp, then it will have the style of life that is characteristic of wasps of that type, whether we know what that style is or not.

The recognition that some individual organism belongs to some type thus carries with it the recognition that the organism has a mode of life, that it dances a dance that traces some pattern or other. This recognition is doubly empirical. First, only empirical investigation of a large number of cases can determine that some organisms fall into some definite type, regardless of the variations that occur among all individuals. Information regarding type is accumulated only through experience, and as an investigator gains more and better experience, he or she comes to have greater knowledge of the details of the kinds of organic life. Griffin knows more about what is to be expected of plovers than I do, and Wooldridge knows more about what is to be expected of *Sphex* wasps than you do. Second, it is an empirical matter to determine just which individuals follow which life patterns. Although natural laws certainly guarantee that there will be similarities between parents and offspring, there is no logical necessity that children will trace the same pattern of life as their parents. So whether an individual instantiates a given type is always a matter for empirical discovery.

The assertion that some organic individual is a token of some organic type carries important implications for what can be expected from that individual. Although there are important variations among the reproductive behaviors of different female *Sphex* wasps, Wooldridge feels justifiable confidence in thinking that he can conduct his research using any member of the same species. In general, the more one knows about the behavioral and metabolic life pattern that characterizes a given type of organism, the more one can infer about the past and future behavior of an individual from the assertion that that individual is a token of that type. The fact that an individual organism has a life process of a certain sort thus has significant explanatory import.

To know that an organism is a token of an organic type is not, of course, to know that the past or future acts of the organism will have certain, definite effects. Not all female *Sphex* wasps in fact change their environments in such a way as to reproduce by building a burrow, and so on. Nor is it the case, in general, that all tokens of the same organic type will do the same things, physically described, in the same external circumstances, physically described. As organic behavior becomes more complex, the relation between stimulus and response becomes increasingly mediated by peculiarities of the internal makeup of the organic agent. One may infer, however, from an individual's tokening of some type that the individual will act at various points in its life history in characteristic ways to achieve the goals that are constitutive of the life of organisms of its type. So, if an insect is a female *Sphex* wasp, and if most of what it does over an interval of time is typical of what other female *Sphex* wasps have done in the process of reproducing, we can infer that this wasp will act in similar ways for the sake of achieving the overall goal of reproducing.

If Fran is such a wasp, and if at this point in the reproductive routine that is typical of this kind of wasp, other wasps generally go into the burrow and drag in the cricket and then lay eggs only if there is no predator in the burrow and the cricket has not been moved, and if this behavior frequently has the effect of preventing the eggs from being eaten, then, when Fran goes into the burrow, she is checking for predators so that her potential eggs will not be eaten.

Conversely, the judgment that Fran is checking for predators by going into the burrow explains why Fran engages in the act of going into the burrow. As developed in the last chapter, to say that S does B for the sake of G implies that

> (1) S is a living agent;
>
> (2) S has some intrinsic goal G*, such that S's doing something at the time it does B that results in G would, given everything else that S's acts result in, tend to realize G*; and
>
> (3) S is an instance of a type of living agent that typically acts so as to realize G* by doing something similar to B that results in G.

So, asserting that what Fran is doing is for the sake of checking the burrow implies that Fran is acting as *Sphex* wasps act when they are at a certain stage in their reproductive routine, a routine that involves their going into the burrow. From these implied facts one can infer the statement that asserts that Fran goes into the burrow at this time. The statement that Fran goes into the burrow at this time is inferred from three general statements and three singular statements, all of which are implicit in the judgment that what Fran does is for the sake of checking for predators. The three general statements are (1) that wasps of a certain type who are engaging in reproductive behavior often go into the burrows they have prepared; (2) that when wasps of this type go into the burrows in this way, their future behavior varies as a function of what is in the burrows; and (3) what wasps of this type do as a function of what is in the burrow frequently results in the prevention of predation of the eggs that the wasp eventually lays. The singular statements are (1) that Fran is a wasp of this type; (2) that she is currently acting as wasps of this type act when they are engaging in their reproductive routine; and (3) that Fran has reached the point in the routine at which wasps of this type go into the burrows they have previously prepared. From all of this, it follows that Fran goes into the burrow.

So saying that Fran goes into the burrow for the sake of checking for predators explains the occurrence of Fran's going into the burrow in a quite straightforward, Hempelian way. The explanation consists in pointing out that Fran is an entity of a certain organic type, that organisms of that type realize a goal typical of organisms of that type, reproduction, by acting in certain ways relative to their environments, and that Fran is currently acting in that way.

Now the three general statements that are implicated in the judgment that Fran went into the burrow for the sake of checking for predators are not exceptionless *laws*, or anything like them, of course. At most, wasps of Fran's type often go into burrows they have previously prepared in the course of their reproductive behavior. There are just too many other factors that can influence wasp behavior to allow for laws regarding whether a given female wasp will go into a burrow that she has prepared in the course of a reproductive routine. Such a wasp might run into a fiendish human experimenter before this point in her routine, for example, and never go near the burrow rather than go into it five hundred times in a row. So there is no possibility of *deducing* that Fran will go into the burrow from the premises implied by the judgment that what Fran does is for the sake of checking for predators. Nevertheless, given those premises, to infer that Fran goes into the burrow is inductively rational. Furthermore, our predictions regarding the future behavior of organisms are much improved when we recognize what kind of organism an individual is and we understand the typical ends of that kind of organism and its habitual ways of attaining those ends. It is perfectly all right to make a trap with only one entrance if you want to catch lobsters, but you had better not try to catch raccoons in such a trap.

Goal-directed explanations thus turn out to be a variety of nondeductive Hempelian explanation. Such goal-directed explanations differ from most Hempelian explanations in an important respect, however. The conditions under which organic agents will act according to general principles include factors internal to the organism that can vary through time, rather than merely external factors. One can infer solely from the fact that some bar is made of iron that it will rust if there is water present in its environment, but not otherwise. But one cannot infer that Fran will go into a burrow that she has prepared solely from the fact that she is a wasp of a certain type that finds herself on the burrow's doorstep. In addition, one needs to know that Fran is at a certain stage of her reproductive routine, that is, that she is acting in order to reproduce.[2] The teleological description of the organism thus plays a crucial role in the specification of the lawlike dispositions mentioned in the explanans.

This dependence of goal-directed explanations on the identification of the particular life stage of an organism creates room for types of errors in teleological judgment that do not occur in nonteleological cases. Judgments of goal-directedness can be false, of course. I can make the judgment that Fran goes into the burrow for the sake of checking for predators even when it isn't true, and there are at least three ways in which I can go wrong. Since the descriptive content of goal-directed judgments implicitly involves the typing of an agent, there can be errors in this typing. Fran might turn out to be a wasp of a different genus, for example, that does not reproduce in this way but rather makes a living by raiding the burrows of its cousins. Alternatively, one can make an error regarding the typical goal-directed behavior of a type of organism, even when one is correct regarding typing. When Europeans first encountered the kookaburra, for example, they were misled by the similarity of the bird's characteristic calls to the sound of human laughter and concluded that the calls were used to express pleasure. Further observation showed, however, that the calls were used to mark territory. That is, the calls occupied the teleological role of keeping other kookaburras away from the vicinity of the caller, and this had the effect of providing the caller with more food, which in turn aided the chances of the individual caller's survival. Sometimes, however, the observer makes no error regarding typing nor any regarding the typical behavior of instances of the type, but still goes wrong. This happens when an individual engages in some piece of behavior that, physically described, is similar to a piece of behavior that at times performs a role in the agent's life process but that on this occasion does not perform that role. One might judge, for example, that some plover that in fact has a broken wing and is limping away from a predator is doing so as part of a broken wing display. In this case, the error is a function of the fact that it is not true that the physical behavior occurs *because* it is an instance of the role that behavior of that physical type sometimes occupies in plover maternal behavior. A plover with a broken wing will limp away from a predator even if she is not a mother, so the counterfactual judgment implied by the teleological assertion that the behavior happens because it is for protecting the young—that the limping would not occur if the eggs were not present—is false. And if it is not true that S does B because it is for the sake of G, then it is not true that S does B for the sake of G.

This covering law–style account of the explanatory power of goal-directed judgments is open to at least two prima facie objections. One concerns the logic of these sorts of explanations as applied to teleological cases. The other reflects general concerns about the kind of biological understanding that underlies this account of goal-directed explanations.

First, this account seems to be closely related to Hempel's account of functional explanations, and there are well-known and devastating objections to Hempel's account that seem to apply to this one as well. Those objections turn on the fact that the existence of some item is never necessary for the performance of some function but is at most sufficient, so that one can never infer the presence of that item from the fact that it performs a function. Similarly, it might seem that there are always multiple ways in which an agent can act for the sake of achieving some end, so that one can never infer the occurrence of a particular act from the fact that the act is for the sake of achieving that goal. So mentioning that that act was for that goal could never explain that act's occurrence.

Second, this understanding of goal-directed explanations takes them to be a species of what Ernst Mayr has called "typological" thinking. Mayr, among others, has argued that the key to the Darwinian revolution in biology was the recognition that this type of explanation needed to be replaced by "population" thinking. Roughly, a style of thought or explanation is typological if events or states of affairs are explained through appeals to the type of the agent or subject of those events or states, together with appeals to other factors. A style of explanation is based on population thinking if the prevalence of a trait within a population is explained by appeal to the distribution of that trait within the population at a previous point, together with appeals to other factors. Clearly, the account of goal-directed explanation that I have offered here is typological, and, one might think, for that reason reactionary.

But there are responses to these objections, responses that will also serve to deepen our understanding of the underlying logic of teleological explanation.

## 2. GOALS AND FUNCTIONS: TYPES AND INDIVIDUALS

According to the view presented here, to say that S does B for the sake of G implies that S is a certain type of organic agent, that agents of that type have a certain intrinsic goal, G*, that they typically accomplish through first realizing G by doing something similar to B, and that B occurs because S is at that stage of its life process at which agents of its type realize G* by first realizing G by doing something similar to B. So, in this view, what goal-directed attributions explain is why the event that has the goal occurs.

This conclusion accords with our ordinary intuitions. But at the same time, there seems to be a problem with this account. All that can be inferred from the premises implicit in the judgment that S is doing B for G is that S does something "similar to B" at the time in question, but as yet I have given no clue regarding the nature of this similarity or why one should be able to infer the apparently singular judgment that B occurs from a premise that mentions only a *type* of individual, "events similar to B."

Essentially, the view I have presented makes the occurrence of B understandable by representing B as a way in which an agent of a definite type achieves a result that it must achieve if it is to continue to exist or is to accomplish some result that is typical of its existence. It is part of the life process of wasps to reproduce. Given the structure and metabolism of wasps, certain proximate results must be achieved if the wasp is to reproduce successfully. In particular, if the wasp is to reproduce, it is necessary that her newborn young be adequately fed and that they be safe from predators. There are characteristic ways in which wasps get this done. Wasps typically reach this proximate end, in part, by going into burrows that they have previously prepared and checking for intruders, reacting differentially depending on what they find. These facts explain why our sample wasp, Fran, does B, that is, why she goes into the burrow in the way she does at the time she does.

Of course, it is not *necessary* that Fran do B to ensure successful reproduction. First, Fran could protect her young from predators by doing something entirely different from B. Indeed, in the experimental situation, what Fran does, B, is not instrumental to her proximate ends at all. Second, even if Fran does something similar to B to accomplish her ends, she need not do B. There are an infinite number of other possible events that would be in their relevant respects similar to B, and not be B. So how can one infer that B occurs from the premises we have mentioned?

These sorts of worries, relating to how a covering law–type of explanation could adequately grasp the nature of goal-directed explanations, are similar to the problems that bedevil Hempel's well-known account of *functional* explanations. Hempel explains the occurrence of a trait i in a system s at time t in the following way:

> (a) At t, s functions adequately in a setting of kind c (characterized by specific internal and external conditions)
>
> (b) s functions adequately in a setting of kind c only if a certain necessary condition, n, is satisfied
>
> (c) If trait i were present in s, then, as an effect, condition n would be satisfied.
>
> (d) (Hence), at t, trait i is present in s.[3]

The standard refutation of Hempel's account of functional explanation is that (d) doesn't follow from (a)–(c).[4] There are always traits other than i that could do n. Clause (c) says only that *if* i were present, n would be satisfied. To reach (d), one needs an *only if*. And this is unwarranted. Let s be a human body operating in normal conditions. Let n be circulating blood and let i be a beating heart. Because an artificial pump also could circulate blood, the circulation doesn't guarantee the presence of the heart.

Does this same refutation apply to the teleological role account of goals? In both cases, what is necessary if the system is to function, or continue, is that, in some initial condition, some end state is achieved. But *how* this state is achieved is left open. Couldn't some event other than B bring about G in the situation in which B occurs, just as some object other than a heart can circulate blood?

We will respond to this question in two stages, corresponding to two different worries about the identity of B. First, while it is true that Fran must somehow accomplish the task of protecting her young from predators if she is to reproduce, it is, of course, also true that there are many ways in which she could accomplish the job of protecting her young from predators, even if we recognize that Fran is a wasp with a certain structure and metabolism. Most of these alternative ways of accomplishing this task are quite dissimilar to what Fran actually does, B. In the functional case, the assumption that an organism with a circulatory system actually will function in a given environment requires only that *something* pump blood, not that this something be a heart. Similarly, the assumption that Fran will reproduce requires only that she do something that protects her young against predators, not that it be going into a burrow at this point. The assumption mandates fulfillment of a condition, not the way in which it is fulfilled. It is the character of the way in which this task is accomplished that we are trying to explain.

All this is true, but it in no way undercuts our analysis of goals. As we have analyzed it, the claim that Fran goes into the burrow for the sake of protecting her young from predators contains both less and more than the analogy with the Hempelian account of functional explanation would suggest. Less, because one can't infer from the judgment that S does B for G that B in fact is sufficient to guarantee that G occur, even when one adds the assumption that S somehow gets G done. Hempel's account of functioning makes the same mistake regarding functions that Wright's account made regarding goals. It simply isn't true that everything that a successful agent does will tend to bring about what it aims for, and it is not true that every item in a functioning system satisfies its function. So our analysis assumes less than does Hempel's regarding the success of the item that is characterized teleologically. On the other hand, our judgment that some agent S does B for G contains an implicit reference to the type of agent that S is. Agents, in our analysis, are typed by reference to a way, or style, of accomplishing the goals intrinsic to life. It is reference to this style that eliminates the possibility that an agent might do something altogether different from what it in fact does to reach the goal of its acts. *Something,* even something metabolically and morphologically similar to Fran, could accomplish the end of protecting her young from predators in some way other than by looking in the burrow; but *Fran,* being the sort of wasp she is, could not. For to be this sort of wasp is, in part, to accomplish this end in this way. Because the claim that Fran goes into the burrow for the sake of checking for predators contains an implicit reference to the sort of wasp Fran is, it requires that Fran protect her young from predators in this way, and not some other.

This response, however, leads to a second possible worry. Because we explain the occurrence of B by appealing to the mode of life typical of the agent's kind, it seems that we can explain, at most, why a kind of event occurs, not that some particular event occurs. And what goal-directed explanations explain is, seemingly, that some particular event occurs.

Goal-directed explanations relate why an event occurs: S does B for G explains why B occurs. But what is the logical character of the designator "B"? There are two possible answers to this question, and the objection turns on choosing the wrong answer. Either B is a proper name, or a demonstrative, that picks out some individual in all possible worlds by rigidly designating only that individual or B picks out an event by giving some description that, more or less narrowly, picks out some event that satisfies that description. If the first possibility is the case, then the objection works. For, of course, showing that there must be some event that is similar to events of going into a prepared burrow that, in other wasps' behavior, bring about safety for young, never could guarantee that just *this* very event, B, must occur. There are always other events, with slightly different spatiotemporal trajectories that would satisfy this description. For example, Fran *could* have moved .01 meters per hour faster and .01 centimeters to the right of where she actually moved and still would have done something similar to the actions of other wasps going into burrows that they had prepared. On the other hand, if B is a description, then the sufficiency of the argument in the explanans to show the necessity of B's occurrence depends on the character of that description.

Consider, yet again, Fran, our model wasp. Fran does something, and we want to explain what she does. There are, of course, many correct descriptions of what Fran does. At least some of these descriptions are adequate to pick out what Fran does from all other actual events in the world, but not from all possible events. Here is one such description: what

Fran did at t was to go into the burrow she had prepared. Let us abbreviate this description as "C." At t, Fran goes in the burrow only once, so C picks out what Fran did from all other actual events. It doesn't, however, pick out what Fran did from all other possible events. Fran could have moved spatially quite differently at t from how she did move and still have gone in the burrow. So C doesn't pick out what Fran did in all possible worlds: it isn't a rigid designator. Let us use "E" as such a rigid designator. Both C and E are possible designators of what Fran did, so they are both possible occupants of the role of B in the explanandum of a goal-directed explanation. That is, we can ask, "Why did C occur?" and we can ask, "Why did E occur?" Now we give a teleological explanation of what Fran did: she did B (what she did) for the sake of checking for predators. Is this explanation sufficient to necessitate that B occur?

Well, that depends. It certainly isn't sufficient to demonstrate the truth of "E occurs." This explanation guarantees only that Fran will do *something* similar to what other wasps of Fran's kind did that resulted in predator avoidance. There are many things that Fran could do, aside from E, that satisfy this description. But is the teleological explanation sufficient to show the truth of "C occurs"? The answer turns on the logical form of event statements, and, in the most widely held views regarding the logical form of such statements, goal-directed explanation *is* sufficient to show this.

There are currently two main metaphysical theories of events, Kim's and Davidson's. According to both theories, events are particulars. And, according to both, when one says that some event has occurred, one is really asserting an existential generalization, almost invariably that there is some event such that it is in some relation to some other individuals. So, when one asserts, "Shem kicked Shaun," one is asserting that there is some event such that it is in the 3 place "kicked" relation, with Shem and Shaun: (Ex) (Kicked(Shem, Shaun, x)).[5] And, when one asserts, "Fran goes into the burrow," one is asserting (Ex) (Going in(Fran, burrow, x)). So, to account for Fran's going into the burrow at t, one must demonstrate that there is at t something such that it is a going into the burrow by Fran. But, given the various parameters involved, to say that Fran does what she does for the sake of protecting her young from predators is to say that, given the type of agent Fran is and what she has been up to, she will do something physically similar to what other wasps have done that has resulted in the protection of their young from predators. But, in this situation, *all* possible acts that satisfy this description are acts that satisfy the description of being a going into the burrow. For the physical similarity in question is precisely that these acts were all acts of going into the burrow that each wasp had prepared. So Fran goes into the burrow. Therefore, if B is read as "going into the burrow," a teleological explanation necessitates that "B occurs" be true.

Thus, if the explananda of goal-directed explanations use rigid designators, the same sort of objection that undermines Hempel's account of functions undermines the teleological role account of goals. If, on the other hand, the events to be explained are designated by descriptions, then, if goal-directed explanations are analyzed according to the model of goal-directed explanations I have presented, the objection fails.

There is very good reason to think that the typical explananda of goal-directed explanations involve descriptions. Indeed, these descriptions are usually pretty vague. We want to know why the chicken crossed the road, or why the plover limped, or why the wasp went in the burrow, or why the frog flicked her tongue. We rarely, if ever, ask why the chicken came to occupy a position with coordinates [x, y, z] by moving in a straight line at velocity v for 4.1 seconds, and we even more rarely ask why some agent did *exactly* what it did. To explain why the chicken *crossed the road,* we need only show that, given the type of agent it is, the current environmental situation, and certain facts about what it has been doing, the chicken does *something* that is a crossing of the road. And this is precisely what a well-constructed teleological explanation shows.

There is, of course, a moral to this result regarding the nature of teleological explanations. When one explains why some nonrational organic agent did something by pointing out that she was acting in order to realize some end, one explains that an event of a certain type occurred. Sometimes the event explained is picked out by a description that itself is teleological. This potential for recursion accounts for the fact that goal-directed explanations are often nested, as we saw above. The wasp builds the burrow (goal-directed description) for the sake of hiding her eggs. But sometimes, and in the most complete goal-directed explanations, goal-directed explanations explain why an event described in physical terms occurs. The wasp, Fran, did move in such a way that at a certain time she went from a position outside the burrow to one inside the burrow, and it is this event that is being explained. But because the event that is explained is accounted for by an appeal to what agents of the same type as the agent of this act typically do, the description used to pick out the event to be explained must be loose enough to cover both this act and many other possible acts that did not occur. Physical descriptions need not be precise descriptions, and goal-directed explanations explain events only under imprecise physical descriptions.

## 3. TYPES AND GOALS: POPULATIONS AND EVOLUTION

According to Ernst Mayr, arguably the most distinguished contemporary American evolutionary biologist,

> the significance of the scientific contribution made by Darwin is threefold:
>
> 1. He presented an overwhelming mass of evidence demonstrating the occurrence of evolution.
>
> 2. He proposed a logical and biologically well-substantiated mechanism that might account for evolutionary change, namely, natural selection. . . .
>
> 3. He replaced typological thinking by population thinking.[6]

As usually understood, the distinction between typological and population thinking is a distinction regarding preferred styles of explanation.[7] Aristotle is thought to be the father of typological thought, and it remained the dominant mode of thought in biology until Darwin. According to the received view on these issues in the history and philosophy of biology, typological thinkers hold two significant theses concerning the members of a group of similar entities, such as a biological species. First, typologists believe that there is some property or set of properties that all and only members of a species possess. Typologists hold that any possible organism that is a member of a species must possess these essential properties. Second, as the term is ordinarily understood, those who practice typological thinking believe that possession of these essential characteristics by members of the species has explanatory significance. So, for a typological mode of thought, the fact that Fran is a *Sphex* wasp implies that Fran possesses certain characteristics, and the fact that Fran possesses these characteristics helps explain why Fran is as she is and does what she does.

Now, according to the currently accepted view of the distinction between typology and population thought, Aristotle and his typological descendants were not dumb. They recognized, as every intelligent person must, that there is a wide diversity among conspecifics with regard to morphology, metabolism, and behavior. Typologists, however, have a distinctive kind of explanation for this variation. Just as there is a natural state that every material body would occupy but for external forces acting on it, so every member of a species would act and be structured in the same way but for factors external to the individual that impeded its natural functioning. (Aristotle was wrong regarding the precise character of the natural state of heavy bodies, of course, thinking as he did that they naturally tended to occupy the center of the earth. Galileo corrected this mistake when he formulated [somewhat inaccurately] the principle of inertia. But this is of no importance, as Galileo retained Aristotle's mode of explanation for the variability of behavior of heavy bodies: all would act in the same way but for the influence of forces that restrained their natural tendency.) So, the variability within a species is a function of the conjunction of two types of forces: the natural tendency of all the individuals in the species to act as their essential nature demands and the tendency of outside forces, such as the material environment and defects in individual structure, to deflect this natural trajectory.

As Elliott Sober has pointed out, this notion of a natural mode of biological functioning that is inherent in membership in a species stands behind our ordinary notions of health and disease, function and dysfunction.[8] The assertion that some individual organism is healthy, or not, is an inherently normative judgment that presupposes a comparison between the way the organism is and the way it ought to be, if it is an organism of that type. A heart that is malformed in such a way that it cannot circulate blood efficiently is just as much a part of the natural world as is a healthy heart. Only when such a heart is compared with a "normal" heart can it be seen as "defective." But the norm in question surely isn't a statistical norm. Rather, it is a norm established by what is natural for a properly functioning human heart. In such cases, we look for the cause of the defect in much the same way we look for the cause of cancer, that is, the factor that deflects the organism from its natural course. So, the medical case displays typological thought insofar as it retains a normative character. Variations in the human population are understood as a function of the way in which a normal situation is deflected by external factors. "Disease" makes sense as a concept in an explanatory scheme that appeals to a norm of correct functioning, characteristic of the members of a species, that would actually occur in those members were the normal course of development not deflected from the outside.

According to Mayr, Darwin saw more clearly than any of his contemporaries the necessity of replacing this way of explaining diversity in biological nature. In population thinking, Darwin's substitute for typology, the diversity within a population is explained by an appeal to the diversity within a *previous* population of the same group, together with the relative likelihood that the individuals in that previous group will reproduce in the environment in which they find themselves and the likelihood of random variations arising across generations. For example, the variation in the prevalence of the sickle-cell trait in different human populations is now seen as a complex function of (1) the variation in the prevalence of the trait in the human population at a previous point; (2) the probability that each of the individuals in that previous population, with their different features, had of surviving and reproducing in the different environments in which they were situated; and (3) the rate of mutation of the sickle-cell trait. That is, the occurrence of sickle-cell anemia within certain subpopulations of human beings is explained by appealing to the likelihood that, within certain environments, individuals with one allele for sickle-cell anemia will reproduce more successfully than will those without such an allele. This increased likelihood of successful reproduction in one subgroup more than compensates for the fact that those with two alleles for sickle-cell anemia are less likely to reproduce than is either of the other subgroups of the subpopulation in that environment. At no point does this explanation appeal to the supposition that it is natural for humans to lack sickle-shaped blood cells or to external factors that deflect those with sickle-cell anemia from their natural course.

We can now formulate an objection to the account of goal-directed explanations that I offered above, one that is based on this distinction between typological and population thinking. The objection has the form of a dilemma. The account offered is clearly a typological explanation: it explains what a biological organism does by mentioning the kind of organism it is and noticing that organisms of this kind normally act in such a way in achieving ends that they characteristically act to achieve. Of course, the proponent of this account recognizes that not every individual of a certain biological type will always act in the same way to achieve its ends. So the explanation isn't deductive in form. Nevertheless, whatever explanatory power this scheme displays depends on the notion that it is normal for conspecifics to act in certain ways, and appeal to normality in explanatory contexts is the hallmark of typological thought.

But typological thought is bankrupt precisely in the biological area that the proponent of this account takes to be the field of teleological judgments. So, either the account is in error regarding how one is to understand teleological explanations in general—and goal-directed explanations in particular—and these explanations must be understood in a way compatible with the population thinking that is typical of modern biology, or the account is correct concerning the structure of goal-directed explanation but, given the bankruptcy of typological explanation, this shows only that goal-directed explanations have no explanatory power.

In a sense, Ruth Millikan grasps the first horn of this dilemma in her evolutionary account of functions.[9] Millikan doesn't deny that an evolutionary understanding of functions fails to capture the entire intuitive content of our ordinary judgments that some item has a function. She does deny that we can make any coherent sense of why functional explanations work as well as they do without appealing to evolutionary history. In essence, her view is that when one says that the function of the heart is to pump blood, one is providing an elliptical explanation of the presence of that item, the heart. One is saying that the heart is there because it pumps blood. Taken at face value, this is a very odd claim. First, just as some act can be for the sake of accomplishing some end without actually tending to bring about that end, so an item can have a function but not the ability to fulfill that function. Witness defective hearts. Second, if one understands the "because" in the explanation in a causal way, as Millikan does, how can the fact that some item has some (possible) effect cause that very item itself?

Millikan's answers to these questions depend on an appeal to the evolutionary history of the item that has the function. The item that has the function of bringing about some result Z is some particular instance of some type of organ, an x. This x is there not because *it* brings about Z in the past, the present, or the future. Rather, there is an item with the characteristics of this item because in the past some ancestor(s) of this x did something that brought about something of type Z. What that ancestor did, of course, brought about a z in the conditions that actually obtained at the time that the ancestor operated. These facts help to explain this x insofar as one can expect that the ancestor will (more or less) accurately reproduce itself (or be accurately reproduced) in favorable circumstances. That it brought about a z in the way it did helps to explain why more copies of that ancestor were reproduced in the area at the time. And these two facts together explain why an entity with the features of x is presently here.

In effect, this reading is designed to solve a hypothetical field functionalist's problem of typing her (perhaps inadequate) specimens by looking at their evolutionary history. One wants to know the function of some item, x. Unfortunately, the item is not doing anything that appears to explain why it is there. Perhaps it is not doing anything much at all. So the field functionalist can't identify the function of this item by appealing to what *it* does. This difficulty is just another implication of the general teleological maxim that if there is something that an individual *should* do, it must be possible that it doesn't do it, so one can't rely on what an individual *does* to identify what it should do. On the other hand, if this item is here because it is a reproduction of some previous item, y, which was subjected to selection pressure, then it is of the type "reproduction of y," even if it is a rather bad copy. If y reproduced more successfully than its competitors because it brought about a z in conditions C, then the function of x is to do what its ancestor did—produce a z in C—even if it never is in conditions C and it never produces a z. That is, the basis of x's having function Z is not what *it* does or even what it could do. Instead, the basis is that x's ancestor, y, did Z; that it did Z explains why x is there. So, whenever an x's being there is explained by appeal to an evolutionary history that has differentially selected x's ancestors for having done Z, then x has the function of doing Z, regardless of whether it can in fact do Z. Alluding to the fact that x has the function of producing Z elliptically explains the fact that x is there: x's being there is caused by the fact that x's ancestors did Z, which is precisely the fact by virtue of which x has the function of doing Z.

Millikan formulates these insights in a precise manner. According to Millikan,

> Where m is a member of a reproductive family R, and R has the reproductively established or Normal character C, m has the function F as a direct proper function iff:
>
> (1) Certain ancestors of m performed F.
>
> (2) In part because there existed a direct causal connection between having the character C and performance of function F in the case of these ancestors of m, C correlated positively with F over a certain set of items S which included these ancestors and other things not having C.
>
> (3) One among the legitimate explanations that can be given of the fact that m exists makes reference to the fact that C correlated positively with F over S, either directly causing reproduction of m or explaining why R was proliferated and hence why m exists.[10]

One might similarly argue that the only biologically respectable account of goal-directedness must appeal to evolutionary history. If S's doing B has the goal of bringing about G, then S must do B because B has the goal of bringing about G. But B need not bring about G for B to have that goal; and even if B does have that goal, it can't be this fact about *B* that brings *it* about. Instead, the fact that Fran's ancestors' going into the burrow caused the eggs of those ancestors to be protected from predators and that this behavior was thus reproduced explains why Fran goes into the burrow. So the fact that Fran's going into the burrow for the sake of protecting her eggs from predators explains her going into the burrow is itself explained by the evolutionary history of Fran and of her behavior.

There are two problems with this way of understanding the implications of Darwinian thought. Both problems have to do with the structure of the explanations offered by population thinking. Millikan, it seems to me, misidentifies both the explanans and the explanandum of evolutionary explanations. Appeals to natural selection explain the *variation* in the range of traits within a population, not the presence of some trait in some individual. Evolutionary explanations account for that variation through an appeal to the relative fitness of some system of which some item was a part, where the notion of fitness depends on the propensity of a system to achieve the goal of reproducing.

Here is the first misunderstanding in Millikan's views. Population thinking accounts for the prevalence of some structure or behavior in some population, not for the presence of some structure or behavior in some individual, as Millikan thinks. One accounts for the fact that a certain percentage of humans have a heart of a certain type by appealing to the facts that some ancestor population had a range of different hearts and that, in the environment in which those ancestors functioned, those with the kind of heart most of us now possess were more fit than the others. The percentages within a population that have a given trait are explained by population genetics; in the most extreme cases, such as that of the heart, such explanations can account for why *almost* all the individuals in a population have some trait. These explanations work by placing the individual trait or kind of organ in the context of the rest of the functional parts of previous organisms and by showing that if the rest of the functional design is constant, individuals with the type of organ in question will be more fit than those with some alternative type of organ. This is why evolutionary explanations explain the preponderance of good design; the good designs, those that tend to survive and reproduce better, predominate. But what is explained is *always* that some percentage of some population, perhaps approaching 100 percent, has such and such features.

But none of these past selection events ever causes some individual to have some trait. The microbiology of particular organisms causes that state of affairs. The defective heart, this very individual with this very structure, is there not because of its ancestor's selection history but because of its causal antecedents in the embryo from which it formed. And that selection history does not explain why *that individual* has the traits it has; the defective heart has those traits *despite* that history. At most, all that one can infer from an evolutionary history that applies population thinking is that a very high percentage of the members of a population will have certain organs that function together in certain ways and that have the overall effect of keeping those organisms alive.

So Millikan is wrong in thinking that appeals to population thinking explain the occurrence of some individual organ or event. But goal-directed explanations *do* attempt to explain the occurrence of particular events. In our standard example, we attempt to explain Fran's going into the burrow on this occasion by pointing out that this act was done for the sake of checking for predators. So evolutionary explanations are always quite distinct from goal-directed or functional explanations. Evolutionary explanations *can* explain why most members of Fran's group go into burrows they have prepared at this point in their reproductive routine. So there is a sense in which Millikan is right in her claim that evolutionary factors explain why teleological explanations work as well as they do. But the sense in which they provide this explanation in no way renders teleological explanations redundant or otiose. Goal-directed explanations are fully compatible with, though not reducible to, the best kind of population thinking in biology, and this disarms the second horn of the dilemma constructed above.

As we have just seen, evolutionary explanations never *directly* explain the facts of individuals' lives and structures. What they directly explain is why certain organs or behaviors are *generally* present within a population, and from this fact one can go on to draw the risky inference that some individual within the population has that structure or exhibits that behavior. Evolution explains why most individuals in some population are what they are and do what they do. What explains, noncausally, why an individual is what it is and does what it does is its membership in a population that mostly is this sort of thing and does these kinds of things. To explain the presence of some organ in an individual, one must infer it from the fact that *most* individuals within the population have that trait. Because the premise is never that *all* individuals in the population have the trait in question, the inference is never deductive, as is witnessed by the example of the malformed heart. Similarly, if one wants to explain why some individual performs some particular piece of behavior, one must infer that explanandum from the fact that most members of the individual's population act in this way in this sort of situation. So the *conclusion* of the evolutionary argument—that most members of a population act in a certain way—is one of the major *premises* implicit in a goal-directed explanation of why some individual in that population acts as she does. In this sense, appeals to population thinking explain why goal-directed explanations work; appeals to evolution explain why the crucial premise in goal-directed explanations is true. But the facts of evolutionary history in no way *cause* the events that goal-directed explanations explain. Such events are caused in the usual way by local physical factors in an organism that acts in conjunction with environmental circumstances. And the adequacy of evolutionary explanations in no way undercuts the need for goal-directed explanations, any more than the adequacy of microbiological causal explanations undercuts the need for evolutionary explanations.

Of course, the fact that an organism shares a wide variety of traits and behaviors with other organisms within the same population in no way *causes* that individual to act or to be structured in those ways. So to the extent that traditional typological thought was committed to an efficient causal role for essences, that tradition was in error. Unlike evolutionary explanations, which explain why a certain distribution of traits occurs among the members of a population, both microbiological and goal-directed explanations attempt to explain why a given event occurs. But only the microbiological explanation provides a cause for the event. The ability to offer teleological explanations for the occurrence of single events nevertheless plays an essential role in the life histories of most animals, including humans.

Consider, once again, Fran's going into the burrow. Why did she do it? She did it for the sake of checking for predators. How does this fact help to explain what Fran did? According to the views presented here, saying that Fran went into the burrow for the sake of checking for predators implies that Fran belongs to a population of wasps, most of whom go into burrows they have prepared in the course of a reproductive routine; that when members of this population successfully reproduce, going into the burrow indirectly brings about protection from predators; and that Fran is currently engaged in just this sort of reproductive routine. From these implied facts, one can inductively infer a high probability of Fran's going into the burrow at approximately this point and thus explain the fact that *Fran goes into the burrow*—the fact that needs to be explained. This is the entire content of the goal-directed explanation. This goal-directed explanation depends crucially on the fact that the predominate mode of life among wasps of Fran's kind involves going into burrows as part of the life process. And evolutionary explanations explain just that sort of fact. They explain why the general premises of goal-directed explanations are often true. But if there were some other explanation for that fact, say God, goal-directed explanations would still work and still have the content that they have. Whatever the reason why wasps of Fran's kind act in these characteristic ways, just so long as they act in these ways, we can provide a goal-directed explanation for why Fran goes into the burrow.

That this should be so makes perfect evolutionary sense, of course. Of vital concern to most individual animals is what other individual animals will do and when they will do it. The ability to predict and explain what an individual will do on the basis of one's knowledge that this individual belongs to some kind and one's knowledge of the habits of that kind—that is, the ability to make goal-directed judgments—is of prime importance for an organism. The knowledge necessary for making these judgments is just the kind of knowledge that might become available in the judge's environment. Vervet monkeys live in an environment that includes both leopards and wildebeests. In general, leopards eat vervets; wildebeests do not. So it is important for vervets to distinguish leopards from wildebeests, even though there may be some leopards that don't like to eat vervets and (less likely) some wildebeests that do. Sometimes, leopards engage in hunting activity; sometimes they don't. So it is important for vervets to distinguish leopards acting on a routine that often results in eating from leopards engaged in other routines, even though leopards sometimes eat opportunistically. When leopards hunt, they often act in characteristic ways. So it is important for vervets to recognize the characteristic pattern of leopard behavior, even though leopards occasionally will do something out of character. But if a vervet is able to make these three distinctions, that vervet is able to act as if it is making judgments regarding that for the sake of which typical leopards act. The ability to make and act on such goal-directed judgments would seem to have an enormous selective value, a value that would guarantee the predominance of that ability in vervet populations.

There are two other issues that are of interest to scientists but of supreme evolutionary indifference to vervets. These are the questions of what causes an individual leopard to act as it does and of why most leopards act in that way. The first of these is properly answered by microbiology, and the second by studies in evolutionary history. But the fact that we can ask these questions that cannot be answered in a teleological manner in no way undercuts the vervet's need to be able to predict what an individual leopard is likely to do next. And this last question is one that the vervet can answer if it is able to make correct judgments about the *goal* of this leopard's current activity.

Our own evolutionary history explains why so many of us are so good at making correct judgments regarding the goals of our fellow animals. Both the content of such judgments and the fact that such judgments are made true or false by real facts about the world, however, are quite independent of that history.

So the fact that goal-directed explanations explain the occurrence of particular events precludes Millikan's attempt to reduce them to evolutionary explanations. But a Millikanesque account of goal-directed explanations faces an even more serious problem. Once one understands how evolutionary explanations actually work, one can see that one *can't* articulate what it is for some behavior or structure to have a goal by appealing to what it was selected for, because what it is for that behavior or structure to have been selected for performing some function can be understood only in terms of a prior understanding of what it is for a behavior or structure to have a goal.

Millikan, in attempting to give an account of what it is for some structure to have a function, helps herself to the notion that the ancestor of this structure did something that explains why this very structure is there or at least explains why such structures have proliferated. That is, Millikan appeals to an evolutionary explanation of the structure in determining the function of the structure. But exactly what is the character of the explanans in such explanations?

One might think that the answer is obvious. In explaining the presence of some trait at some time or the prevalence of some trait within some population at some time, one might look to the comparative fitness of the ancestors of that trait at some previous time relative to the fitness of other, similar traits in the population at that previous time. If the ancestor trait had a higher fitness value than any of its competitors, then this relatively higher value explains, through natural selection, the presence or prevalence of its descendent. This answer is essentially correct, but it is also far more complex than is usually recognized, at least by most philosophers.

As Susan Mills and John Beattie pointed out in their classic article, the fitness of an individual can't be identified with the actual number of offspring of that individual, and the fitness of a phenotype can't be identified with the actual number of offspring produced by individuals with that phenotype.[11] Such identifications would render the notion of fitness explanatorily empty. Question: Why does individual A produce more offspring than individual B? Answer: Because A is fitter than B. Question: What is it for A to be fitter than B? Answer: For A to produce more offspring than B.

Because actual reproductive success cannot establish fitness, to say that A is fitter than B must mean something like "The phenotype that A realizes has a higher *propensity* to produce offspring than the phenotype that B realizes." But what does "propensity" signify here? The relative propensity to reproduce can't be identified with actual reproduction, even in the long run. Accidents can and do take a toll on organisms that are fitter than their rivals. Instead, as Beattie and Mills should make clear, the "propensity to produce offspring," and thus the fitness of a phenotype, must be measured in terms of optimality models. "Evolutionary biologists frequently *derive* relative fitness claims from optimality models; this is basically an engineering design problem. It involves determining, solely on the basis of design considerations, which of a set of alternate phenotypes maximizes expected descendant contribution."[12] But what is an optimality or, as John Maynard-Smith calls it, an optimization model?

Here is how it is supposed to work: First, assume that there is some definite set of possible phenotypes that are candidates for selection. Second, assume that there is an end (or, as Maynard-Smith would have it, an "aim" or "goal")[13] to be accomplished or maximized. Ask which of the set of phenotypes has the best design for accomplishing that assumed goal. Then determine whether that phenotype is actually displayed in succeeding generations. If it is, one's model has been confirmed; if it is not, one must alter one's hypotheses in some way. Confirmation can be quantitative: if one's model can specify mathematically how much better than the alternatives a given phenotype should be at accomplishing the hypothesized goal, one can predict the prevalence of a phenotype within a population. When the model is confirmed, one can look at the role that some particular trait or behavior performs in reaching the hypothesized goal in the hypothesized phenotypic design, and from that information one can conclude that the trait is present or prevalent in the actual population because it was better at performing that role—it was more fit for doing that job—than its competitors and was selected to perform that function. That is, for some trait to have been selected for performing some function is for it to have contributed to the propensity of some phenotype for achieving some goal by being suited to achieve some subordinate end corresponding to its "function," which is specified by the design specifications of the phenotype in which it operated.

That aim or goal might seem obvious, but it is not. The nature of the goal is problematic in two interrelated ways. First, the assumed aim or goal to be maximized by a design, *is part of the hypothesis being tested.*

> Three sets of assumptions underlie an optimization model. First, there is an assumption about the kinds of phenotypes or strategies possible (i.e., a "phenotype set"). Second, there is an assumption about what is being maximized; ideally this should be the inclusive fitness of the individual, but often one must be satisfied with some component of fitness . . . . In testing the optimization model one is testing the adequacy of these hypotheses to account for the evolution of the particular structures or patterns of behavior under study.[14]

To be sure, Maynard-Smith makes the general assumption that it is inclusive fitness (that is, the propensity to reproduce that is common to all individuals of a similar genotype) that is maximized. But this is *just* an assumption. As George Williams, himself no friend of group selection, makes clear, the *only* way to tell whether it is the phenotypical fitness of a set of individuals, or the fitness of a group, or inclusive fitness, that is maximized by natural selection is to construct optimization models that assume that these various goals are maximized and to determine which of them best match the observed structures prevalent in a group.

> A direct assessment of the importance of group selection would have to be based on an accurate knowledge of rates of genetic change, due to different causes, within populations; rates of proliferation and extinction of populations and larger groups . . . etc. . . . Obviously this ideal will not be met, and some indirect method of evaluation will be necessary. The only method that I can conceive of as being reliable is an examination of the

> adaptations of animals and plants to determine the nature of the goals for which they are designed. The details of the strategy being employed will furnish indications of the purpose of its employment. I can conceive of only two ultimate purposes as being indicated, genic survival and group survival. All other kinds of survival . . . will be of the nature of tactics employed in the grand strategy, and such tactics will be employed only when they do, in fact, contribute to the realization of a more general goal.[15]

This leads to the second problematic aspect of the way in which "goal" is used here. Consider the issue of group, as opposed to genic, selection. According to both Williams and his opponents, it is an empirical issue whether some trait or behavior is prevalent in a population because it serves the goal of reproducing that *population* or because it serves the goal of reproducing the *individuals* in that population. But in either case, the individuals that make up the population themselves count as reproducing. So it doesn't follow from the facts that some individuals are the product of reproduction and that the traits of those individuals are the product of natural selection, that those traits are themselves the product of the fact that the ancestors of those traits served the goal of reproducing the ancestors of those individuals better than their competitors. That is, the goal that some trait of some individual serves cannot simply be identified with the role that this trait played in explaining the reproductive capacity of the ancestors of that individual. This lack of identity renders problematic the very notion of the goals or purposes for which an adaptation was designed.

The references to "design" and "engineering" considerations in the biological literature suggest that the proper place to look for an account of the use of "goal" or "aim" in these discussions is Robert Cummins's classic article on functional analysis. But this suggestion is, at best, misleading. Central to Cummins's analysis is the principle that a functional analysis of a system explains a disposition or capacity of a system by appealing to the dispositions or capacities of the component parts of the system and the way in which those parts are linked together in the system. But which disposition such a functional analysis helps to explain is not part of the explanation, it is the condition that is to be explained. In evolutionary explanations that appeal to optimality models, on the other hand, the end or goal served by a product of natural selection *is part of the hypothesis to be tested* and thus part of the explanation. As we just saw, this goal cannot simply be identified with the reproduction of some individual or some phenotype. Instead, that some trait served some goal in a prior generation helps to explain the prevalence of that trait in a later generation, but what that goal is, or what it is for some trait to serve some goal, is not thereby identified or explained. That is, evolutionary explanations of the functions of structures or the goals of behavior presuppose an understanding of what it is for a structure or behavior to serve a goal, but do not themselves provide an account of what it is for a structure or behavior to serve a goal.

So what does all of this have to do with Millikan? Just this. Millikan attempts to explicate the teleological notion of a function by appealing to what she takes to be the unproblematic notion of what an item with a function was selected to do by natural selection. If the above is correct, then *that* some item was selected for satisfying some function implies that it contributed to the goal of maximizing the fitness of some containing system. But which containing system that is, and what it is for an item to serve some goal, is not thereby defined. In that case, to explicate the notion of a goal by appealing to a Millikanesque, evolutionary account would be to explain in a circle; one would be explicating what it is for some present structure or behavior to serve a goal by appealing to the goal it served in the past. And the detour via selection history supplies no additional insight. That is, one *can't* articulate what it is for some behavior or structure to have a goal by appealing to what it was selected for, because what it is for that behavior or structure to have been selected for performing some function can be understood only in terms of a prior understanding of what it is for a behavior or structure to have a goal.

CHAPTER FIVE

# INSTRUMENTAL RATIONALITY

The behavior of living organisms aims to achieve goals, even when those organisms do not act rationally and have no beliefs or desires. When a female wasp goes into a burrow she has prepared, her act sometimes has the goal of checking for predators, even though, as Wooldridge demonstrates, the wasp does not act rationally and has no beliefs or desires.

But among animals that act in order to achieve goals, some, such as human beings, do act rationally. Because both human beings and wasps act to achieve goals and because human beings, but not wasps, act rationally to attain their ends, it seems that we should classify rational action as a species of goal-directed behavior. But what distinguishes rational action as a special class of goal-directed behavior?

There is an intimate connection between the concepts of belief and desire, on the one hand, and the concept of rational action, on the other. When animals act rationally to achieve their goals, they do what they do because they have states with intentional contents, such as beliefs and desires, which provide reasons for what they do. The wasp does not act rationally and does not have any beliefs or desires of her own, whereas we humans both act rationally and have beliefs and desires that provide reasons for what we do.

It is no accident that those living beings that act rationally are those living agents that have beliefs and desires that motivate their behavior. As we will see, rational action essentially involves versatile adaptive behavior that varies in response to changing circumstances, responds effectively to the source of the agent's mistakes, and is novel in relation to the agent's species-specific pattern of life. Such rational action can be interpreted and explained teleologically only if (1) the agent has the intrinsic goals typical of its species *and* (2) the agent has unique, changing, and idiosyncratic goals and perceptions of the world, states that need not be shared with other members of its species. I will argue that these unique, changing characteristics of rational agents are necessary to explain their novel ways of coping in a rationally adaptive fashion with a changing environment and form the core of our notions of belief, desire, and intentional content. My second task is to begin to articulate a new account of what it is for an agent to have beliefs and desires with intentional content, an account that depends on the roles of those kinds of states in explaining rational action.

## 1. THE TELEOLOGICAL STANCE AND THE INTENTIONAL STANCE

For an organism to be alive, it must have certain characteristic goals that it achieves in certain characteristic ways. As life involves a constant interchange between the living thing and its changing environment, what the organism must do in order to achieve its goals varies as a function of the determinate state of its environment. All organisms, insofar as they are alive, act differently in different environments. When an organism fails to respond appropriately to its environment—that is, when it fails to do what it must in order to achieve its primary goal of staying alive—it ceases to be alive. It dies.

These facts have led Daniel Dennett to conclude that all living things, and many nonliving things, such as thermostats, can correctly be said to have beliefs and desires that explain their behavior. "A thermostat, . . . I claim, is one of the simplest, most rudimentary, least interesting systems that should be included in the class of believers—the class of intentional systems, to use my term. Why? Because it has a rudimentary goal or desire (which is set, dictatorially, by the thermostat's owner, of course), which it acts on appropriately whenever it believes (thanks to a sensor of one sort or another) that its desire is unfulfilled."[1]

Dennett's suggestion is not loony. It simply trades on a remarkable set of facts. In teleological terms, an agent's act has both a goal and a set of circumstances under which that act would succeed in attaining that goal. Insofar as the agent stays alive, its acts succeed most of the time in attaining their goals, and so most of the time what it does is just what it needs to do to achieve its goal in the actual circumstances. What the *Sphex* wasp does when she goes into the burrow has the goal of checking for predators so as to protect potential young. And, on many occasions, what a wasp does when she goes into the burrow has the effect of protecting her young from predators in the circumstances in which she acts.

Beliefs and desires are states of an agent that together rationally explain how that agent acts. These two types of states are individuated by their *contents:* one desire is different from another because it is the desire *that* p, while the other is the desire *that* q; one belief is different from another because it is the belief *that* p, while the other is the belief *that* q. It is one thing for me to desire that I eat ice cream, and quite another for me to desire that I eat cream cheese. It is one thing for me to believe that there is ice cream in the freezer, and quite another for me to believe that there is cream cheese in the freezer.

But all desires are such that their respective contents correspond exactly with the *goal* of any act that they help to explain. If I act on my desire to eat cream cheese, then what I do has the goal of my eating cream cheese; if I act on my desire to eat ice cream, then what I do has the goal of my eating ice cream. The contents of a belief specify the conditions under which an act that the belief partially explains would succeed in reaching its goal. If I act on my desire for ice cream and my belief that there is ice cream in the freezer, then, all else equal (and much has to be equal), if my belief is true, then the act that these states explain will be successful in achieving its goal of my eating ice cream.

Given these correspondences and the facts that every organism acts with goals and, insofar as it continues to live, acts mostly appropriately in actual conditions, every organism *can* be interpreted as acting rationally, just as Dennett says. The formula is simple. Take the overall goals of the organism; notice the typical sequence of discrete steps that it takes to achieve those goals and the conditions under which those discrete acts actually succeed; then ascribe beliefs and desires with the corresponding contents to the organism. The agent desires those states of affairs that correspond both to its overall goals and to each of the intermediate goals it achieves in the course of achieving its overall goals. The agent believes, truly, that the states of affairs that obtain when it acts successfully are actual; when it acts inappropriately in the actual circumstances to achieve its goals—which happens frequently—the agent believes, falsely, that the conditions in which its act *would* succeed are actual. If one can assign an overall goal to thermostats, one can even assign beliefs and desires to them that explain their behavior.

The relationships that underlie Dennett's "intentional stance" account of intentional content (between the contents of beliefs and desires, on the one hand, and the goals and environmental conditions for successful goal-directed action, on the other), will also underlie the account of intentional content that is developed in the following chapters. Desires, in general, are states of agents that explain why the agent's acts have the goals they do, and the contents of the desires that an agent acts on just are the goals of the acts that they help to explain. Beliefs, in general, are states of agents that explain why agents act as if certain conditions actually obtain when those agents act to achieve goals. The contents of the beliefs that an agent acts on just are the conditions under which their goal-directed acts would be successful.

But there are at least two major problems with Dennett's use of these facts. As we have already seen, there is a difference in principle between entities that have ends of their own—living organisms—and entities such as thermostats, to which one may assign purposes extrinsically. Only the former act in a goal-directed fashion.

More importantly, even when "physical stance" and "design stance" explanations are not available, much organic behavior simply requires no rational explanation. In these cases, simple goal-directed explanations will do, and such explanations make no use of the supposed intentional states of agents. Dennett has missed the "teleological stance." Furthermore, in the cases in which the teleological stance is appropriate, the beliefs and desires that would need to be assigned to the agent to explain its behavior are not related to *each other* as they would need to be for the agent to have genuine reasons for acting as it does. The behavior of the *Sphex* wasp simply isn't *rational* enough to support the inferential connections among beliefs and desires that are constitutive of the very notion of an agent that acts for reasons, and thus constitutive of the very notion of an agent that genuinely has beliefs and desires.

But the behavior of some other organic agents, even when in the service of simple organic goals of those agents, cannot be entirely explained without appeals to transient intentional states of those agents. There is a difference between the behavior of a wasp and that of, say, a human being. The behavior of the human is instrumentally rational, while that of the wasp is not. This difference corresponds exactly to the difference between those agents that have beliefs and desires and those that don't. Beliefs and desires, and, in general, intentional mental states with contents, are essentially *reasons*, and agents have reasons only when they act rationally.

## 2. REASONS

To this point, I have argued that some behavior is directed toward a goal even though the agent of that behavior lacks beliefs and desires that account for what the agent does. Because what it is for an act to have a goal makes no reference to the beliefs and desires of agents, there is no reason to think that all goal-directed acts are the result of an agent's content-laden mental states. It is possible to make perfectly good nonmetaphorical sense of the claim that a wasp, or a plant, or a bacterium, acts for the sake of achieving a goal without assigning beliefs or desires to such organisms. One can explain why these organisms act as they do by appealing to the goals of these acts, without assuming that these agents act because of their beliefs and desires.

It is possible, then, that some organic agents act in order to achieve goals but lack mental states with intentional content that help to explain how they act. But surely some organisms, such as human beings, both act for the sake of achieving ends and have intentional mental states, such as beliefs and desires, that account for how they act. Moreover, the goals that creatures such as we act to achieve are themselves indexed by our intentional mental states. The goals of our acts covary with the contents of our desires; in general, you or I act in order to realize goal G only if we *want* G. How we behave to achieve our goal or satisfy our desire covaries with the contents of our beliefs; in general, all else equal, if you want to protect your children and believe that you can do so by checking the wiring in your house, you check the wiring.

Organic agents that act because of their beliefs and desires act in order to achieve ends. This fact is rooted in the logic of rational explanation. Beliefs explain actions only when they are combined with desires, and when an agent acts out of a desire, what that agent does has the goal of satisfying that desire. If I want *that my children be protected* and act on that desire, then the goal of what I thereby do is *that my children be protected.* In organisms that act because of their intentional mental states, the content of the desire that partially explains their act coincides exactly with the goal of the act.[2]

When an agent acts because of her intentional mental states, the content of the desire that partially accounts for her act fixes the goal of the act. The content of the relevant belief plays a different role. The content of the belief fixes *what exactly* the agent does to satisfy her desire. Because actions will bring about different results in different circumstances, the action required to achieve the goal of an act varies as a function of the circumstances in which the act occurs. But having a goal doesn't guarantee achievement of that goal, even for acts that result from the intentional states of their agents. So what the intentional agent does in detail is not simply a function of her desires and her environment. What is done in detail is a function of the agent's desire and what the agent *takes* her environment to be like; that is, a function of what she desires and what she believes. If I act on my desire to protect my children and believe that *my* checking the wiring will protect them, then, all else equal, in order to protect my children *I* will check the wiring. If I believe that I am not competent to check the wiring but that someone else is, then, all else equal, in order to protect my children I will arrange for that person to check the wiring.

The principle of charity popularized by W. V. Quine and Donald Davidson holds that most of an agent's beliefs must be true. We can now see the teleological facts that underlie this principle. For an organic agent to act for a goal, it is necessary that the agent continue to live. For the agent to stay alive, most of what the agent does must be successful at attaining its goal. All agents that act rationally are organisms, and, thus, most of what they do must be successful at achieving its goal. Since the contents of the beliefs of rational agents covary with the conditions under which an act would be successful at achieving its goal and since most of what the agent does must succeed in the actual conditions, the contents of beliefs must correspond to actual conditions most of the time. That is, the beliefs of any rational agent—that is, any agent that has beliefs at all—must be true most of the time.

The contents of the intentional mental states of an agent are relevant to what that agent does in a quite unique and extraordinary way. One cannot appeal to the contents of the intentional states of an agent to explain the agent's acts unless the factors appealed to in the explanation *rationalize* or *justify* the act. Let's say that I jumped up and down and that what motivated this act was my desire to protect my children and my belief that my checking the wiring would be instrumental to doing this. According to some views regarding the nature of belief and desire, this is a possible scenario: a person could be so wired that events that are tokens of the belief and desire in question could also cause a person to jump up and down and, in that sense, motivate the jumping. But in that case, the intentional *contents* of these states would be irrelevant to the explanation, and the fact that these states were beliefs and desires would play no role in the explanation. These contents could play no role because they are irrelevant to the justification of the jumping and thus fail to rationalize the jumping. They are irrelevant because a justification of action always involves a special relation among the contents of the motivating belief and desire and the act they motivate, a relation that is absent here. If a belief and desire rationalize or justify an act, then, were the belief true, the act would *in fact* bring about the desired state of affairs. But in this case, jumping up and down would not bring about my children's being protected, even if my belief that checking the wiring would help protect my children were true. So the contents of this belief and this desire do not justify the jumping and thus are irrelevant to the explanation of the jumping.

When the goal-directed acts of an agent are explained by appeal to the contents of the agent's intentional mental states, we say that the agent has *reasons* for what she does and that the agent exercises *instrumental rationality.* An agent has reasons for what she does just in case her behavior displays rationality of an instrumental sort.

The behavior of a *Sphex* wasp has a goal that is similar to the goal I have when I check the wiring. We both act with the goal of protecting our young. But I act rationally, and the *Sphex* wasp doesn't. Because of this difference in our behavior, we account for what the wasp does and for what I do in different ways. We explain my behavior by mentioning my reasons, which explain what I do through justifying and rationalizing what I do. We know that such rationalization and justification is relevant to the explanation of what I do because what I do is instrumentally rational. We account for the *Sphex* wasp's behavior by situating it within behavior typical of wasps of her type and without mentioning any distinctive states of the wasp herself, with or without intentional contents. Although there are good reasons for what the *Sphex* wasp does, the wasp herself has no reasons to act as she does: she simply acts in that way. To have reasons, she would need to have beliefs and desires that explain and justify her acts, and she has no such intentional states. We know that she has no such states because she does not behave rationally.

The difference between those agents that act in order to achieve goals and have intentional states with content and those agents that act to achieve goals and lack such states is the difference between those agents that act in an instrumentally rational fashion and those that don't. Only agents that act rationally have reasons of their own for what they do. And only agents that have such reasons act as they do because of the intentional contents of their beliefs and desires. We can know that an agent has intentional states with contents only if that agent acts as she does because of the contents of those intentional states. So our knowledge of the intentional states of agents derives entirely from our understanding of the agents' behavior as rational. Given these facts, if one wants to know what it is for an agent to have intentional states that have contents, it would be wise to attempt to understand how instrumentally rational behavior is different from goal-directed behavior that is not rational. Or, to put the issue more graphically, it would be well for us to understand how an instrumentally rational agent, such as a human being, differs from a *Sphex* wasp.

First, however, a word of caution. I will argue in later chapters that instrumental rationality is not the only rationality; there are other kinds that are not identical with or reducible to instrumental reason. To be a person involves actions that display these other forms of rationality, as well. So the difference between an agent, such as a *Sphex* wasp, that acts in a goal-directed but not instrumentally rational fashion and one that acts in an instrumentally rational fashion is not the whole difference between wasps and people. But no entity can have any mental states with content unless it acts for reasons, and it cannot act for reasons unless its behavior displays instrumental rationality. So no agents can display these other forms of reason unless they are also instrumentally rational. The difference we are about to explore, that between goal-directed but nonrational agents—such as wasps—and instrumentally rational agents—such as vervets, humans, and perhaps some birds—is thus a crucial distinction, corresponding as it does to the distinction between the classes of agents that have no intentional mental states and those that do.

## 3. NONRATIONAL BEHAVIOR: THE LESSONS OF THE *SPHEX* WASP

The difference between agents that have reasons for what they do and those that don't, is not the difference between agents that act in order to realize goals and those that don't. What the *Sphex* wasp does has the goal of protecting her young, just as what I do sometimes has that goal. Nor is it the difference between more successful and less successful behavior. Given the prevalence of wasps, it is a good bet that they are at least as good at protecting their young as I am. It is a difference between two styles, or ways, of acting to achieve goals.

These styles differ because the ways in which various sorts of organisms respond to their environments also differ. All goal-directed organic behavior is reliably responsive to local differences in the environment of the agent. Insofar as an organism remains alive, it manages to do what it must to stay alive, and as the environment varies, what the organism does to reach this goal must also vary. Since what a living thing does to stay alive varies with the environment, and since what the organism does depends on what the organism *is,* all organisms must be able to sense changes in their local environment. This is why even Dennett's thermostat must be equipped with sensors. Every organism engages in a linked series of acts that have the goals of bringing about a linked sequence of results in the changing environments in which they occur, a linked series that follows the sequence of proximate goals typical of the organism's kind of life. Life itself is a process, and every organism has some species-specific form of life. So the acts of every organism have both proximate and overall goals.

But organic beings that display instrumental rationality act in ways appropriate to achieve their goals in a variety of changing circumstances that they cannot sense, and they act for proximate goals that are not fixed by their species-defined life processes. This difference allows instrumentally rational agents to respond successfully to alterations or details in their environments for which no routine response is programmed in the life pattern of the species to which they belong. By inferring the presence of conditions they cannot sense from those they can and by developing sequences of acts with novel goals that are capable of ultimately achieving their fixed, overall, organic ends in these novel conditions, instrumentally rational agents can hope to cope with novelty and obscurity. Only such instrumentally rational agents, who alter the goals of their immediate acts and alter their acts so as to succeed in novel circumstances, can properly be said to have reasons for what they do.

Let me illustrate the difference graphically. A *Sphex* wasp is generally very good at reproducing, as long as there are no scientists around to move their crickets. But as soon as something unique, unexpected, or simply incapable of sensible differentiation by wasps turns up in the environment, the individual wasp is at a loss.[3] Instrumentally rational agents can deal with such occurrences, however. Because rational agents can infer beliefs regarding the presence of nonsensible conditions and alter their behavioral patterns so as to act for novel proximate goals, such agents can hope to reach their ultimate ends, come what may.

We can illustrate this difference in style between rational and nonrational agents by recalling the biological description of the behavior of the *Sphex* wasp.

> When the time comes for egg laying, the wasp *Sphex* builds a burrow for the purpose and seeks out a cricket which she stings in such a way as to paralyze but not kill it. She drags the cricket into the burrow, lays her eggs alongside, closes the burrow, then flies away, never to return. In due course, the eggs hatch and the wasp grubs feed off the paralyzed cricket, which has not decayed, having been kept in the wasp equivalent of deep freeze. To the human mind, such an elaborately organized and seemingly purposeful routine conveys a convincing flavor of logic and thoughtfulness—until more details are examined. For example, the Wasp's routine is to bring the paralyzed cricket to the burrow, leave it on the threshold, go inside to see that all is well, emerge, and then drag the cricket in. If the cricket is moved a

> few inches away while the wasp is inside making her preliminary inspection, the wasp, on emerging from the burrow, will bring the cricket back to the threshold, but not inside, and then will repeat the preparatory procedure of entering the burrow to see that everything is all right.[4]

From Dean Wooldridge's perspective, this behavior is interesting because it *appears* to be rational but can be convincingly shown to be nonrational. If we can isolate those aspects of the wasp's behavior that give the appearance of rationality from those that conclusively show that the wasp is not rational, we can infer the conditions necessary for an agent's behavior to count as instrumentally rational.

Wooldridge reports several details of behavior that sustain the illusion of the wasp's logic and thoughtfulness, that is, sustain the appearance of instrumental rationality. Prominent among these details is the way in which the pattern of behavior is "elaborately organized" and exquisitely adjusted to bring about just the end that marks the goal of the behavior. This pattern is not merely minimally capable of bringing about this end. In many circumstances, it would bring about this end in a nearly optimal way. For example, were the cricket killed by the sting rather than paralyzed, much of the wasp's behavior could remain the same, and a fair percentage of healthy young would result much of the time. But the fact that the cricket is paralyzed rather than killed serves to optimize the chances of a high percentage of extremely healthy young in most circumstances. Similarly, the wasp goes into the burrow "to see that all is well." Although Wooldridge is silent on the topic, this description makes sense only if the wasp varies her subsequent behavior depending on what she finds in the burrow, laying eggs if there is nothing inside or nothing has been moved and not doing so if either has occurred. One can imagine a fairly successful wasp that skipped this part of her reproductive routine. But any wasp that engages in this searching behavior thereby optimizes the chances of success. And these elaborately organized nuances of optimizing behavior sustain the illusion of rationality.

Why does the fact that the wasp paralyzes rather than kills the cricket seem to support the judgment that the wasp is acting rationally? If the wasp were acting rationally, then her actions would correlate with the content of her beliefs and desires. Either killing or paralyzing the cricket serves the general goal of supplying the young with meat. But the goal served by paralysis *rather than* killing is *fresh* meat for the young. There is no fact in the immediate and local environment to which the wasp is responding when she acts by paralyzing rather than killing. The explanation for this differential behavior must be in the wasp. So we are tempted to attribute the corresponding desire to the wasp: she wants her young to have fresh meat. But if the wasp indeed acts on this desire, then the companion belief that motivates what the wasp does must be something like "This cricket will still be fresh to eat if I sting it in this way but not in that." But how could the wasp know or come to believe *that?* Nothing within her local environment contains that information. So, *we* conclude, *the wasp* must have *inferred* this belief from other beliefs that she had in virtue of her placement within the environment. And all agents that act on such inferred beliefs are rational and have reasons for what they do: so the wasp is rational.

The wasp's checking behavior adds to the illusion of rationality by seeming to support the suggestion that some of the wasp's beliefs are inferred. The significance of the checking behavior for the success of the wasp's overall goal relates to a certain possibility. The wasp's checking the burrow seems to further the goal of having healthy and happy young because *if* the wasp were in position to act differently were there something dangerous in the burrow (as opposed to the way she acts when there is nothing dangerous there), *then* she would be more likely to succeed in her overall project. That is, the goal of this activity seems to be to gain information regarding which of a range of possibilities concerning potential danger are actual. So it seems as though the wasp has the desire to gain this information. But, in that case, the belief that would work with this desire to explain the behavior would have the content "I could find out whether there is something dangerous in this burrow if I checked inside."

There are two distinct aspects of this belief/desire pair that give the appearance of rationality. First, since there is nothing in the actual environment that generally covaries with the presence of the behavior, the fact about the wasp that explains the behavior can't be assigned a content more definite than "I could find out whether there is *something* dangerous in this burrow if I checked inside." And second, since there is nothing in the actual local environment that the wasp is responding to, the wasp seems to act on the belief that she *could* find out *whether* there is something dangerous in the burrow. But how could the wasp come to have *this* belief? An agent can find out that an individual is present in its environment if it has sensors capable of registering this fact. But what sensation could ever register the fact that or account for the belief that "I could find out whether there is something dangerous in the burrow if I checked inside"? The content of this belief involves a *possibility* that there could be something described only as "dangerous" in the burrow. This belief is not keyed to any actual fact in the wasp's environment. So, we conclude, the wasp must have *inferred* this belief from other beliefs based solely on actual interactions with her environment.

When we treat the wasp's act of stinging the cricket in such a way as to paralyze it rather than kill it and her act of going into the burrow as if they were the result of beliefs and desires, the beliefs that the wasp would need to act on could be inferred by the wasp only from information actually present in the environment. We suppose that this inferred belief is the source of the wasp's behavior because we can think of no actual environmental factors that could signal to her that these acts are appropriate in the circumstances for reaching her evident goal, reproducing. The wasp approaches its end goal by first acting to achieve something else, that is, gaining information or paralyzing the cricket rather than killing it. When the wasp is seen as rational, she is also seen as having these proximate goals *because* she has the beliefs that we take her to have inferred. We think that it is only *because* the wasp believes that she could better serve her potential young if she had information regarding the contents of the burrow, and she wants to protect her young, that she acts so as to gain this information by going into the burrow. The appearance of rationality is also the appearance of the agent's capacity to establish new proximate goals for its acts on the basis of inferred beliefs. When an agent can do this, we say that it acts on its desires, for desires are essentially continuing but contingent states of an individual agent whose contents explain the goals of this agent that are not species-specific or species-determined.

But, alas, the *Sphex* wasp is not rational, as Wooldridge informs us. We know she is not rational because of her failure to respond appropriately to the scientist's intervention. Such an intervention is a novel environmental factor for which a pattern of life could have no fixed and established response. If the wasp were rational and her acts were based on an ability to infer beliefs from information available to her in the environment and an ability to establish new proximate goals in light of those beliefs, then the wasp should be able to respond to the scientist's intervention with appropriate behavior that would serve the ultimate goal of protecting her young. For example, the wasp could act so as to gather information from the area around the burrow to help her to discover a cause for the movement of the cricket. This is exactly what one would expect from the wasp if her earlier "checking" behavior in the burrow were the result of rational processes. Any being capable of inferring that it was valuable to her enterprises to gather information in the one case ought to be able to infer from that success that it was valuable to gather information in the other. But the wasp does not act with the goal of gathering information in the experimental case, and, given the inferential linkages, this failure indicates by modus tollens that she did not infer the belief that it would be valuable to gather information in the natural case, or act for reasons when she went into the burrow.

When confronted with the odd fact that the cricket has moved, the wasp merely moves it back and repeats her earlier checking routine, although none of this behavior is instrumental to her ultimate purpose. This routine shows that the wasp was always acting according to a set species-defining behavioral pattern and that the appearance of rationality was just an illusion. *Natural selection* has equipped this wasp, as it has equipped most wasps, with a behavioral pattern that optimizes the chances of reproductive success in most of the circumstances in which wasps typically find themselves. But natural selection has not equipped this or any other wasp with the ability to respond to novel conditions in the environment by generating new proximate goals that are instrumental to her ultimate ends or with the ability to infer from what she can sense in the environment how she should act to attain these new goals. These are the abilities necessary for instrumentally rational action.

The wasp has no beliefs, because one has beliefs only if one can infer beliefs, and to have a belief one must also, in general, have the beliefs that can be correctly inferred from it. And the wasp has no desires, because to have a desire one must be capable of acting in order to achieve novel proximate goals on the basis of one's beliefs, and the proximate goal of each of the wasp's acts is fixed by the behavioral pattern distinctive of this type of wasp.

## 4. INSTRUMENTALLY RATIONAL BEHAVIOR: THE LESSONS OF THE PLOVER

How is a plover different from a *Sphex* wasp? What is it about the behavior of birds that arouses the suspicion that they have reasons for what they do? Consider, again, Donald Griffin's description of the plover's broken wing display.

> Ornithologists and ethologists have repeatedly observed the behavior of nesting plover when a large intruder . . . approaches a nest where a killdeer or piping plover is incubating its eggs. At a considerable distance, . . . the plover may stand up and walk slowly to a point a few meters from the nest. . . . The bird may then walk rapidly or fly in almost any direction except toward the nest. If a person approaches these birds while feeding or when they have no eggs or young, they fly away from the intruder to a safe distance. . . . The bird flutters slowly but conspicuously away from the nest, staying relatively close to the intruder. . . . If the bird is walking or running, its gait is different from normal locomotion and more noticeable. . . . Often the tail almost drags on the ground, and the wings are slightly extended, sometimes one more than another, strongly suggesting some weakness or injury. After running a few meters, the bird may flop about on the ground, extending one or both wings, as if injured. This is often called the "broken wing display," and it requires considerable effort for an observer to believe that the bird is really quite healthy. . . . Throughout most of this predator-distraction behavior, the bird watches the intruder. Typically it does not move in a straight line and stops from time to time. If the intruder approaches, the bird moves farther ahead. If not, the bird usually flies back closer to the intruder and repeats the behavior. . . . Typically the bird continues the injury simulation while leading the intruder some distance away from the nest or young. Finally, however, it flies away rapidly, usually in the same direction, then circling back to the general vicinity, though seldom to the exact spot, where the eggs or young are located.[5]

There are several notable similarities between the behavior of the plover and that of the wasp. Most significant, of course, is the fact that both the bird and the insect engage in a quite complex pattern of activity that has the goal of providing for the safety and well being of their young. In the course of this complex behavior, both the bird and the insect take actions that have the goal of ensuring that the young are not devoured by predators. On the other hand, there is a marked difference between the complexity of the plover's behavior in response to the potential danger from predators and the simplicity of the wasp's response. The wasp goes into the burrow prior to laying the eggs and responds differentially depending on the presence or absence of disturbance. But, as the experimental situation makes clear, her differential responses are limited to a very small range of options and are highly stereotyped.

The plover, on the other hand, engages in a series of acts that involves a high level of sensitivity to various behaviors by the potential predator. There are numerous behavioral forks in the plover's activity at which the bird responds differentially in a way so sensitive to the behavior of the intruder that one is tempted to believe that she is responding to her estimation of the intruder's intentional psychological state. For example, the way in which the bird moves—hopping away from the intruder, dragging her wing, if she sees the intruder moving toward her, or flying toward the intruder, if she sees the intruder not following her—tempts one to say that the plover responds differentially depending on whether she believes that the intruder has *noticed* her display.

Beyond such details, however, it seems that the plover would engage in the display only if she had an estimation of the contents of the psychological states of the intruder. The goal of the display is surely to induce the intruder to move away from the nest and thereby to protect the young. But if the plover desires this proximate end, then it would also seem that she desires another more proximate end, that of leading the intruder to believe that she is injured. The intruder follows the mother plover, after all, only because the intruder believes that the plover is injured and thus an easy potential meal, a premise it has reason to believe only because of the display. Because the display succeeds in achieving either of its proximate ends only if the display in fact causes the intruder to believe that the plover is injured, it appears that the bird *must* believe that if she acts as if she is injured the intruder will come to believe that she is and that this belief gives her a reason to act as she does.

This apparent responsiveness of the plover's behavior to the potential psychological states of the intruder imparts to the behavior its convincing flavor of logic and thoughtfulness. For the ability to respond appropriately in light of the intentional states of other agents is a paradigmatic type of rational behavior. Intentional mental states, such as beliefs and desires, are essentially factors appealed to in the course of rationally explaining the goal-directed behavior of agents. As such, there is no way for one agent to sense the presence of such states in another agent.[6] So if an agent acts in a way that can be successful only if some other agent has some particular intentional psychological state, it seems that there is no option but to assume that the first agent has *inferred* the presence of that state from the evidence that is sensibly available to it in the behavior of the other. Any agent that acts on inferred beliefs regarding imperceptible present conditions in fact has, and acts on, beliefs that count as the agent's reasons. That is, such agents are instrumentally rational.

Thus the apparent sensitivity of the plover's behavior to the intentional psychological states of the intruder provides the most convincing evidence that the plover is acting rationally. There is nothing comparable in the behavior of the *Sphex* wasp. Nevertheless, this apparent ability of the plover to respond appropriately to intentional states in other agents does not settle the issue of whether the plover is really acting in an instrumentally rational fashion, regardless of Griffin's enthusiasm. It is possible that our reasons for thinking that the plover has reasons for what she does could be neutralized by further observation, just as the details of the wasp's behavior under experimental stress undercut the appearance of rationality. And the details of how this could happen serve to sharpen our sense of the necessary behavioral conditions on instrumental rational activity.

It appears that all nesting female plovers will go into the broken wing display if any intruder of the right size approaches the nest. It also seems that any female plover will enact virtually the same script in carrying out the display: If the intruder approaches, drag wing behind and move away from intruder and nest. If the intruder does not approach, fly toward intruder and leave cover. And so on. These facts should be enough to rouse suspicion, for they suggest the possibility that the plover, just as much as the wasp, is not acting rationally at all but merely instantiating a preset pattern of behavior.

The experimental situation reveals both that the individual wasp is enacting a predetermined species-general script triggered by certain physical factors and that the wasp is incapable of generating effective novel behavior to attain her ends by inferring true beliefs regarding how to proceed from the sensory information available to her. These two facts are closely related. If the wasp were capable of generating novel effective behavior, that behavior would, by its novelty and effectiveness, cease to be an instance of a stereotyped species-general response to a physical situation. Rather, it would be a response appropriate to the significance of the particular circumstances in which the wasp found itself. To test whether the plover is acting for reasons when she engages in the broken wing display, then, we need evidence regarding the degree to which a plover is capable of generating behavior appropriate to shifting circumstances, behavior that is not part of the standard, species-general plover repertoire.

One could test this, for example, by seeing how a plover responds to the introduction of a new sort of intruder, one that acts only in pairs. It would be relatively easy for a pair of experimenters to act in concert, so that one of them approaches the nest, but only if the other has followed the broken wing display of the mother away from the nest. In such a case, the broken wing display would become counterproductive to the usual goal of keeping intruders from the nest, as neither of the intruders would ever go to the nest if the bird did not offer the display.

It seems highly unlikely that a plover could ever develop the appropriate novel response to this situation. It doesn't follow from this, of course, that plovers entirely lack rationality or reasons for what they do, as rationality is something that comes in degrees, and one can be more or less rational. Such an experimental outcome would demonstrate only that plovers are less rational than the scientists who study them. But we knew that already.

This sort of thought experiment does demonstrate, however, that rationality demands the ability to *learn* to perform new tricks in aid of one's biologically determined ends. If plovers, like wasps, are entirely incapable of generating new effective goal-directed behavior on the basis of their sensory experience of the actual situations in which they find themselves, then we are justified in asserting that plover behavior is as nonrational as that of wasps and that plovers have no beliefs and desires because they have no reasons for what they do. The supreme test of instrumental rationality is the ability to learn to respond effectively to a range of situations by engaging in novel, but goal- and situation-appropriate, activity.

This ability to learn is the *behavioral* mark of instrumental rationality par excellence: it is the only behavioral mark that reliably indicates that an agent is generating new proximate goals and inferring beliefs, and the ability to do these things is the minimal condition on counting as having reasons for what one does.

There are, in fact, excellent reasons to believe that at least some birds are pretty rational in this sense. Consider the very well-documented case of the British tits that learned both how to drink cream—which they certainly were not evolved to drink—and how to open British milk bottles.

> Behavioral versatility came into play in a spectacular fashion in the 1930s when two species of tits discovered that milk bottles delivered to British doorsteps could be a source of food. At that time milk bottle tops were made of soft metal foil, and the milk was not homogenized, so the cream rose to the top of the bottle. One or more birds discovered that the same type of behavior used to get at insects hidden under tree bark could also be used to get cream from milk bottles. . . . A change in the technology of covering milk bottles eventually ended the whole business, but meanwhile thousands of birds had learned, almost certainly through observation, to exploit a newly available food source.[7]

The British milk industry came up with effective countermeasures, so tits are not as rational as people. But the flexibility and novelty of the behavior of the tits that first applied the techniques for opening milk bottle tops and the perspicuous observation of the tits that learned this technique, are surely powerful evidence that at least some birds are capable of acting rationally.

## 5. GOALS AND REASONS

The *behavioral* mark of instrumental rationality is novelty. Instrumentally rational agents are not merely capable of doing things that neither they themselves nor any other of their kind have ever done, although they are indeed capable of that kind of innovation. Instrumentally rational agents are also able to act for proximate goals that no other of their kind has ever acted to attain. Any wasp has not only fixed ultimate goals (survival, reproduction, and so on) but also fixed *means* of attaining those goals. Indeed, in an agent as nonrational as a wasp, the distinction between the goal of an activity and the means of achieving that goal tends to evaporate. A means of attaining a goal is just something that an agent accomplishes in the course of achieving an ultimate end, but is not part of that end itself. Life is a process that involves a sequence of coordinated repeating steps. And much of what the wasp does in the course of staying alive can be seen indifferently either as a means to survival or as part of or constitutive of the process of being alive. Just as a mammal breathes so as to continue living, but breathing is also part of what it is for that mammal to be alive, a female *Sphex* wasp builds a burrow to reproduce, burrow-building is part of the life history that every female *Sphex* wasp instantiates insofar as she lives as a female *Sphex* wasp.

In instrumentally rational agents, the lockstep between the stereotypical pattern of the life of the agent and the means the agent employs to continue that pattern falls apart. The agent *is* alive, so the agent has the goals intrinsic to its type of life. These goals include not only such universals as survival and reproduction, abstractly described, but also more specific ends constitutive of the biological pattern of the agent's life. Animals that are biologically equipped to eat only meat must act with the goal of finding meat to eat, whether they are rational or not. These determinate goals constitute fixed teleological signposts that constrain the interpretation of behavior; all else equal, an agent's acts should be interpreted as having the ultimate goal of achieving one or another of these biologically fixed ends. In nonrational agents, these fixed signposts exhaust the goals of the acts of the agent. The goal of a wasp's going into a burrow she has prepared is fixed by the role of that act in the life history of *Sphex* wasps, and the wasp never does anything that has a goal that is not fixed by such a role. But because instrumentally rational agents are capable of novel and teleologically successful responses to novel and unique environmental conditions, much of what they do that serves to achieve their fixed goals cannot be seen as instantiations of the life patterns of their species and cannot be predicted or explained through appeal to those patterns.

Nevertheless, such rational and, hence, mostly successful behavior does fit into a sequence of acts that achieve one or another of the fixed ends of the agent by way of doing something else first. That is, the instrumentally rational agent achieves its fixed biological goal *through* first making it a goal to accomplish something else and then acting to accomplish this something else. This "something else," this proximate goal of the novel rational behavior of an instrumentally rational agent, is the means the agent uses to achieve its ultimate goals, and such an agent acts with the proximate goal of actualizing the means to its ends.

Most of what such instrumentally rational agents do in the course of attempting to achieve their novel proximate goals must be appropriate to attaining those goals, and most of the proximate goals themselves must fit together in a way appropriate to attaining the agent's biologically fixed ultimate goals, on pain of the agent's no longer counting as rational. We introduce rational explanation only when we encounter a goal-directed agent that is *very good* at achieving its goals in *very many* situations. Unless an agent generally does what it should to achieve its goals in very many situations, it does not count as rational. This is the first normative fact about rational agents: in general they do what they should, given their ultimate goals, in lots of different and previously unencountered environments.

This normative fact points the way toward the resolution of a tension that the attentive reader will surely have noticed in this account. On the one hand, I have argued in previous chapters that what it is for an act to have a goal involves a reference to the species-general behavioral pattern characteristic of the agent's type of life. On the other hand, I have argued in this chapter that instrumentally rational action is behavior in which an agent acts for proximate goals that are not typical of the behavior of its species. But, then, how can it be that such behavior has any goal at all?

The resolution of this apparent dilemma is straightforward. Reference to the characteristic behavior of the members of a species is necessary in any specification of the goals of the acts of nonrational agents because not all goal-directed acts succeed. In the cases in which such acts fail, there is a difference between what the act actually brings about and what it should have brought about, and, since it is necessary to the normative dimension of goal-directedness that such a distinction be possible, there must be some way to fix the counterfactual condition that counts as the success that serves as the standard for measuring failure. This standard can be fixed only by recognizing that the agent is a living organism of a certain type and, as such, acts in certain characteristic ways in order to achieve certain intrinsic ends.

But this fact is as true of *rational* agents as it is of nonrational agents. Tits are avian organisms that need to breathe air if they are to stay alive, that have certain sorts of digestive systems and hence must ingest only a certain range of organic matter if they are to stay alive, and that come equipped with beaks but not teeth or fingers. They also come equipped with an intrinsic set of ultimate goals and a fixed pattern of proximate goals that they must achieve in the course of reaching those ultimate goals. But in addition, tits do some novel things that don't *duplicate* this fixed pattern but rather *fit into the pattern* as a subordinate part. The seemingly rational behavior of the tit is, as it were, an epicycle or curlicue attached to the grand cycle of the tit's life. The individual tit that first came up with the trick of opening milk bottles and drinking the cream took resources available in the tit's morphological and behavioral repertoire and applied them to achieve an end that was also already fixed by *what it is to be a tit,* that is, nourishing itself by ingesting and digesting a certain range of animal protein.

The ultimate goal of instrumentally rational animal behavior is always an already characteristic end general to the biological type of the agent. We recognize an agent as rational only if we repeatedly and consistently find that the agent successfully does novel things in new environments that indirectly lead to the attainment of its already specified biological ends. If, and only if, an agent acts in this way do we feel ourselves compelled to explain what the agent does in the course of these novel, if indirect, behavioral sequences by reference to transient factors present in the individual agent herself: the agent's beliefs and desires. It is the *success* of the novel behavior in achieving ends that we already know the agent has that drives us to the ascription of these states. For beliefs and desires explain why the instrumentally rational agent frequently acts in the actual situation in such a way as to achieve indirectly its ultimate ends.

It is for this reason that beliefs and desires have contents. A desire is something about an agent that accounts for why it acts with the proximate goal that it does. A belief is something about the agent that accounts for why it acts in detail to achieve its proximate ends in the (mostly successful) way it does. The formula for ascribing such states to agents that act in an instrumentally rational way is just a more detailed development of the formula that stands behind Dennett's intentional stance: (1) Ascribe to the agent the desire to achieve its ultimate goal; (2) ascribe to the agent desires whose contents correspond to the various stages that it actualizes as means when it successfully reaches its ultimate goal; (3) ascribe to the agent those beliefs whose contents correspond with the actual conditions under which the instrumental acts of the agent actually succeed in getting what the agent wants, given (2); and (4) when an agent doesn't succeed in attaining one of its biologically fixed ends, but would succeed were conditions somewhat different, ascribe to the agent the desires to attain the intermediate stages that would have led to this goal had conditions been different and those beliefs that would have been true were those conditions actual.

Because instrumentally rational agents are rational, however, we will not need to invoke the fourth clause so frequently that we have no grounds for distinguishing the successes of instrumentally rational activity from its failures . In the overwhelming majority of cases, instrumentally rational agents desire what they should, given the ultimate ends they are working toward and the means they are using to attain them, and they believe what they should in order to achieve those ends in these ways. In effect, the pattern of mostly rational *reasons* for the individual agent's acting as she does to achieve biologically established ends takes the place of the reference to the way in which successful members of the agent's species achieve those ends. It is by reference to what the agent *should* have done to achieve the ultimate goal that we know she has, that we determine what she was *trying* to do and how she was trying to do it.

So an instrumentally rational agent can have a novel proximate goal for some aspect of its behavior just to the extent that it is rational. It is the rationality of the behavior in pursuit of the biological ends of the agent that allows for the possibility of the behavior's having novel proximate goals at all. But such agents have desires just to the extent that they have novel proximate goals—desires being, essentially, states of the agent that we use to explain the agent's novel proximate goals. Agents have beliefs just to the extent that they have desires. For beliefs are factors specific to agents, factors that account for the details of the way in which instrumentally rational agents act to achieve their novel proximate goals.

From these results one can infer two significant conclusions. First, only agents who have goals that are *not* fixed by desires can have goals that *are* fixed by desires. For only instrumentally rational agents have desires that fix goals, and only agents that act instrumentally and indirectly in novel ways to achieve goals that are fixed biologically (rather than by the agent's desires) can be instrumentally rational. Second, the contents of an instrumentally rational agent's beliefs and desires depend on the roles played by the acts that those states explain in the rational goal-directed behavior of the rational agent. Even though beliefs and desires explain what it is the agent does, the *content* of the beliefs and desires depends on the fact that those states rationally explain the goal-directed acts that they do. As it turns out, the fact that instrumentally rational agents mostly do what they should to attain their biological ends in very many novel situations has important implications for what sorts of beliefs and desires such agents can have.

## 6. THE NORMS OF INSTRUMENTAL RATIONALITY

By the very nature of the case, any instrumentally rational agent will act for proximate goals that are not programmed into all members of its species. Why such an agent acts for the goal it does, then, cannot be because it is a biological agent of the type that it is. The most significant fact about the way the members of a rational species attain the goals that they must in order to stay alive is that they have been equipped by evolution with the capacity to generate new and appropriate proximate goals in response to novel environmental problems: there is no guarantee that every, or even most, members of the same rational species will come up with the same solutions to those problems.

Since we can't account for the fact that an instrumentally rational agent acts for the proximate goals she does by appealing to her fixed biological nature, we must explain it by appealing to some unique feature in the individual agent herself. We say of such an agent that she "desires" to attain her proximate end. We introduce the desires of a particular agent to explain why she acts to attain some novel proximate goal or other. The tit wants to open the milk bottle. Different instrumentally rational agents of the same species will want to attain different proximate ends; only some tits want to open milk bottles. Because all *Sphex* wasps act with similar proximate goals, there is little reason to think that the wasp wants to gain information.

What an instrumentally rational agent does to achieve its proximate ends varies in different conditions. Since the proximate goals of instrumentally rational agents can vary among members of the same species, one can't account for what, in particular, such an agent does to attain its proximate ends by an appeal to what any agent of the same species would do to attain the general ends of that species. Rather, what explains the actions of the particular agent is some feature of that agent itself. We say that such an instrumentally rational agent acts as she does because of the content of her "beliefs." We introduce the beliefs of a particular agent to explain why she acts as she does to attain the proximate goal of her acts. The tit pecks at the bottle top because she believes that doing so will gain her access to the source of nourishment inside. Different instrumentally rational agents of the same species will believe different things about how to attain their ends; only some tits believe that pecking bottle tops will gain them access to nourishment.

As rational behavior, most of what an instrumentally rational agent does to achieve its ultimate goals must be successful. But since rational agents achieve their goals by employing appropriate means and by doing something else first, this implies that the acts of such agents must be related in certain ways. The proximate goals of the acts must fit together in such a way that, were they severally successful, the agent could attain its ultimate biological goal. Since instrumentally rational agents mostly do what they should in actual novel circumstances to achieve their ultimate goals, most of what the agent does to achieve its proximate goals must be successful in the actual circumstances. That is, most of what such an agent does must be appropriate in the actual conditions under which it is done to achieve its proximate goal, and those proximate goals must be so related to each other that when one is attained, the agent is capable of subsequent behavior which is successful in fixing and achieving further goals that lead to the ultimate attainment of the biologically fixed goals of the agent. The fact that some British tit could differentially pay attention to milk bottles with soft foil tops would give us no reason to think that it was instrumentally rational unless it could go on to open those tops, and follow that up with drinking the cream and thereby succeed in nourishing itself.

In nonrational agents, the goal of a particular act is fixed by the relation between the given act and other acts of the agent, in light of the pattern of action that is general to the biological type of the agent. The biological type of the agent cannot fix the pattern against which the goals of the particular acts of a rational agent are typed, however. Instead, for the individual instrumental acts of a rational agent to count as having goals at all, they must fit together in such a way as to count as rational. To repeat the minimal statement of what is involved in that pattern, most of what such an agent does must be appropriate in the actual conditions under which it is done to achieve its proximate goal, and those proximate goals must be so related to each other that when one is attained, the agent is capable of subsequent behavior which is successful in fixing and achieving further goals that lead to the attainment of the biologically fixed goals of the agent. This pattern of relations among the instrumental acts of rational agents provides the necessary context for assigning a goal to each of the instrumentally rational agent's acts.

Since the contents of the desires of instrumentally rational agents covary with the proximate goals of the acts that they help to explain, the fact that the goals of the acts of an instrumentally rational agent must be related in a certain way also constrains the way in which such an agent's desires must be related, if the agent is to count as rational. Since the contents of the beliefs of instrumentally rational agents covary with the conditions under which their acts would successfully achieve their goals, the facts that those acts must mostly be successful, and must be related to each other in a certain way, constrain both the relations among the contents of the beliefs of such agents and the relations between those beliefs and the actual environment. That is, the normative fact that instrumentally rational agents mostly do what they should to attain their ultimate ends in many different novel situations by developing appropriate and novel means places a series of normative constraints on the contents of the beliefs and desires of these agents.

The explanation that we introduce to account for the facility of rational agents for dealing with the world hypothesizes that there is something about the individual agent that accounts for that facility, an ability to generate "beliefs" regarding the environment which dispose the agent to act as it should regardless of the character of the environment. That is, when we suppose that an agent acts on reasons we suppose that it acts as it does because it believes that if it acts in *this* way it will change *this* environment so as to achieve its goal. And since rational agents in general act as they should in lots of different environments, this supposed ability of rational agents to produce beliefs is also a supposed ability to produce the preconditions for correct behavior for attaining goals in lots of different environments. If the agent couldn't do this, it wouldn't be rational. But this constitutes a second normative fact about rational agents: in general, rational agents believe what they should in order to attain their goals in many different environments.

Since rational agents are successful in a wide range of environments, they are likely to be successful in environments that at first don't seem very promising for organisms of their organic type. Human animals need water and oxygen and food and warmth and so on, and none of these things are to be found one hundred miles straight up. And yet, as I write this, there are human animals living one hundred miles straight up. To make this possible, a whole host of linked alterations must be made in the natural world. The necessary alterations are linked in the sense that the ultimate conditions that make human life possible in space depend on previous alterations that in turn could not occur unless even more previous alterations had occurred. Rational agents are successful much of the time in making these linked alterations: otherwise they would not count as successful in coping in a wide range of novel environments, as they must in order to be rational. But these prior alterations in the environment must themselves be brought about by goal-directed activity, and there is no goal-directed activity without a goal. It is essential that rational agents, as rational agents, act for the goals that they desire, that is, that they be motivated in part by their desires. So rational agents must desire the proximate goals that are necessary to achieve their ultimate, organically prescribed, goals, on pain of not being rational. This is another normative fact about rational agents: rational agents, in general, desire what they should in order to achieve the ultimate objects of their desire. This is a *constitutive* fact about instrumental rationality. That rational agents usually satisfy Kant's dictum that to will the end is to will the necessary means is a necessary truth about instrumentally rational agents.

Rational agents respond appropriately in a wide range of situations, and they often do so by engaging in a linked sequence of acts that together transform the natural environment into one in which they can more directly attain their ends. To act successfully in this way, rational agents must desire what they ought to desire in order to attain their ultimate, biologically specified ends. But this isn't enough. The ability to generate novel desires in response to new situations would be of no help to an agent in reaching its goals if that agent were not also able to generate novel, and correct, beliefs. This generation of beliefs involves two distinct abilities. As we have already seen, one of these is the ability to generate, in new situations—whether created by its own acts or not—the beliefs that the agent should have in those situations if it is to attain its goals. If an agent were not able to adjust its behavior to the new circumstances generated by its previous acts, then it would be unable to make use of the results of its previous activity in its subsequent activity. So any rational agent must, in general, be able to generate the beliefs that it should given its environment.

But agents also need a second belief-generating ability if they are to count as engaging in rational action. The linked behavior that the rational agent uses to transform its environment involves the ability to desire what that agent should so as to attain its ultimate end. But what it should desire is relative to its ability to further transform the environment created by the earlier acts in a future sequence so as to reach its goal. So insofar as an agent desires what it should, given its ultimate end, it also believes what it should about what it could do, were it to attain its proximate ends. Otherwise, it could not engage in the long-range successful projects typical of rational beings. Let's say that I want to get to work and that I have misplaced my car keys. As rational, I desire the necessary means to my ends—that is, I desire what I should given my ultimate ends and, given the circumstances, I should desire to find my keys. Given that I want to get to work, it is rational for me to want to find my car keys, however, only if I can use those keys as part of a sequence of acts that would get me to work. So I should desire to find my keys only if my beliefs are such that, were I to find them, I could get to work—that is, if I believed as I should regarding how I should act were the currently counterfactual circumstance of my having my keys made actual.

This amounts to a fourth normative fact about agents engaged in rational activity: in general, rational agents believe what they should regarding how they could attain the ends of their desire, were they to find themselves in currently counterfactual circumstances. This in turn amounts to a requirement that, in general, rational agents be able to generate the instrumental beliefs that they should regarding counterfactual conditions. This is a belief-generating faculty separate from the ability to generate the beliefs that they should, given the actual environment. The first, more primitive mechanism amounts to an ability to accurately sense an environment. Insofar as animal life demands differential responses to different significant environments, any animal, rational or otherwise, must be able to sense what it should about its real environment. But such sensation amounts to the ability to generate beliefs that can function as reasons only if it is linked with a second faculty, the ability to generate the beliefs that one should about nonactual circumstances. There is no way that such beliefs can be generated from immediate interaction with the agent's actual environment.

Where, then, could these appropriate beliefs come from? The holistic character of the attribution of reasons supplies the only possible answer. Just as rational agents do what they should, given their beliefs and desires, so they also *believe* what they should, even regarding counterfactual circumstances. We explain the correctness of rational action by appealing to the beliefs and desires that count as the reasons for that action. Reasons explain an action insofar as they motivate and justify that action. We explain the fact that the rational agent comes to believe what it should about nonactual environments in just the same way, by appealing to her reasons. One has reason to come to believe something just in case one has states that, taken together, justify coming to have that belief, and one's reasons explain coming to have that belief if they motivate coming to have that belief. Some agents—rational agents—believe what they do because their other beliefs justify their coming to have those beliefs. An agent is justified in coming to have some belief, B, if it believes something else, A, such that if the agent's acting as if the content of A corresponded to actual conditions would lead to successful action, then the agent's acting as if the content of B corresponded to actual conditions would also lead to successful action. When an agent comes to hold a belief that is justified in this way, it is appropriate to say that the agent has *inferred* its new belief.

Now, rational agents in general believe what they should about what acts would be successful in nonactual environments. That they believe as they should concerning these matters is explained by their inferring these beliefs from their other beliefs and desires. So we now can see that there is a fifth normative fact about rational agents: in general, rational agents infer what they should infer from their beliefs and desires.

This final principle rounds out our normative description of rational agents. To recapitulate, rational agency satisfies five constraints: (1) In general, rational agents do what they should, given their goals, in lots of different and previously unencountered environments; (2) in general, rational agents believe what they should in order to attain their goals in many different environments; (3) in general, rational agents desire what they should in order to achieve the ultimate objects of their desire, their organic ends; (4) in general, rational agents believe what they should regarding how they could attain the ends of their desire, were they to find themselves in currently counterfactual circumstances; and (5) in general, rational agents infer what they should infer from their beliefs and desires.

These principles are part of a normative description of rational agency. The principles are descriptive in that it is a necessary truth about any agent that acts rationally that it satisfies these principles. For these principles are constitutive of what we mean when we say that an agent acts rationally, and we explain what an agent does by an appeal to its reasons only when the agent acts as these principles prescribe. On the other hand, when an agent satisfies these principles, she is satisfying a norm. The agent is acting, desiring, and believing as it should. What marks these principles as norms of *instrumental* rationality is that this "should" is grounded in the goals or ends of the agent, treated as a state of affairs to be attained through the actions of the agent. The pattern of action, belief, and desire that we attribute to an agent that acts rationally to attain its goals is a pattern that is constituted by the satisfaction of these norms. The rational agent acts, believes, and desires as it should in order to get what it ultimately needs, given its biological character.

# CHAPTER SIX

# RATIONAL EXPLANATION

## 1. JUSTIFICATION AND EXPLANATION

Both rational animals and nonrational animals act in order to attain goals. Both the wasp and the plover act with the goal of protecting their young, and there are good reasons for them to act as they do. There are good reasons for the plover to attempt to deceive an intruder by engaging in the broken wing display, of course. But because a predator might have discovered the burrow, there are also good reasons for the wasp to go inside to see that all is well. Only the plover, however, might possibly act as she does *because* of those reasons; that is, only the plover might have reasons of her own for acting as she does.

The explanation for why the wasp goes into the burrow turns on the character of the wasp as an animal of a certain type. We explain what the wasp does by saying that she does it in order to see that all is well. This style of teleological explanation, however, involves an elliptical way of alluding to the fact that this wasp is an organism of a certain type, with certain definite intrinsic goals that are typically attained, when they are attained, through a stereotypical sequence of behaviors. That this wasp engages in this type of act is no accident. This wasp has reached a point in her reproductive routine at which wasps of her type typically go into a prepared burrow and, depending on what is inside, act differently. In cases in which the wasp reproduces, this differential behavior ensures that predators do not eat the prepared cricket or the wasp eggs. Knowledge of these facts allows one to predict and explain what this wasp will do in this situation, and all of this is elliptically contained in the judgment that the wasp goes into the burrow to see that all is well. But the factors that teleologically explain why this particular wasp acts as she does have to do with the sort of wasp she is, not with anything peculiar to this wasp.

When we explain what a rational agent does by alluding to that agent's reasons, we use the same form of speech as when we teleologically explain the actions of the nonrational wasp. We say that the (hypothetically rational) plover goes into the broken wing display in order to deceive the intruder and thereby to protect her young. But we are not appealing to exactly the same range of facts in each sort of explanation. It is central to the notion of rational agency that what the individual rational agent does is a function of facts that are specific to that individual. Rational agents are agents capable of acting for proximate goals that no other agent of the agent's type has ever acted to attain before. The first tit that attempted to peck its way through the top of a British milk bottle acted with a goal that no other tit had ever attempted to achieve. Because of this, no appeal to any behavioral pattern typical of tits could ever explain why this tit acted in this way. Because the goal-directed behavior of this tit is unique, the teleological explanation of this behavior must also refer to facts unique to this bird.

This particular genius of a tit is still a tit, however, and the fact that this bird is a tit, and thus an animal of a certain sort, *is* relevant to the teleological explanation of this tit's act, even if this fact cannot possibly fully explain what the tit does. Tits are animals, and, as animals, all tits have certain intrinsic goals, including the goal of surviving. Furthermore, tits are animals of a certain definite sort with a definite type of morphology and a limited, though versatile, range of behaviors that they can use to accomplish their intrinsic goals. In particular, all tits, including our rational genius, have beaks but no teeth or fingers, and all tits use their beaks to remove obstacles in the way of their reaching food sources necessary for their survival. These facts, together with the further fact that many of the tits that follow our genius's example actually drink the cream in the bottles and thus ingest food that helps to nourish them and keep them alive, allow for a teleological interpretation of this genius tit's novel behavior. All tits have the goal, among others, of feeding themselves, and all living tits sometimes achieve this goal by removing obstacles to food sources with their beaks. We know that the cream in the bottle is a potential food source for this tit and that this tit is using its beak as all tits do when they are removing obstacles from their access to food sources, although this tit is removing a novel obstacle to a novel food source. So this tit's behavior *would* be instrumental to achieving one of this tit's goals, nourishing itself, were the behavior successful in removing the top of the milk bottle. So, we conclude, what this tit does when it pecks at the bottle top has the goal of opening the top so that the tit can drink the contents, even if the tit fails to achieve this end. And the further facts that this tit actually succeeds in drinking the cream in the bottle and that its fellows follow up on this success with similar successful acts of their own serve only to confirm us in this teleological judgment.

In cases of simple, basic, instrumentally rational action, it is thus possible to assign proximate goals to novel behavior even when such behavior is not part of a specieswide pattern of behavior. In effect, one isolates the novel sequences of behavior that in fact achieve the biologically fixed intrinsic goals of the agent in novel and unique ways. One then breaks down those sequences into their instrumentally significant parts and assigns goals to these acts in accordance with the role each plays in attaining the ultimate ends of the agent. The pecking is for the sake of opening the bottle, which is for the sake of drinking the contents, which is for the sake of nourishing the tit.

In this context, it is important to note that the point—the goal—of the novel behavior of rational agents can be specified—indeed, can make sense—only against the background supplied by the fixed, intrinsic goals and behavioral and structural resources characteristic of the agent's biological type. The goal of an action is a holistic property of that act and involves relations among that act, other acts of the same agent, and the agent itself. What it is for our rational tit to want to open the bottle, and what it is for the tit to act in order to open the bottle, *presupposes* that the tit intrinsically has the goal of nourishing itself in order to stay alive. Those interpretationist philosophers (to use Daniel Dennett's phrase) who attempt to analyze what it is for a rational agent to have beliefs and desires in terms of a teleological theory (Donald Davidson's term) of an agent's beliefs, desires, and actions *always* presuppose such a teleological biological context, although they almost never recognize this presupposition or acknowledge its importance and necessity.[1]

But *why* does our rational tit act with the goal of opening the bottle at this time? *That* it does so can be discerned by placing the act in biological context. But what *explains* why it acts in this way with this goal? The answer, of course, is that the tit acts as it does because it has beliefs and desires which give reasons for acting in this way. That is, we think that there is a rational explanation for her behavior.

In the case of simple goal-directed behavior, the nonrational agent acts with the proximate goal that it does because it is an agent of a certain sort that has reached a certain point in a stereotypical routine of achieving one of her intrinsic goals. One mobilizes these facts in offering the explanation that the agent acts as it does because it is acting to attain some goal. But it can't be that the inventive tit acts as it does because it has reached a certain point in its feeding routine and because all tits that reach this point act in this way. This tit acts with a proximate goal, the goal of opening the milk bottle, a goal that is unique in all of titdom. So the judgment that this tit acts with the goal of opening the bottle can't include the information that all tits act in this way at this point in their feeding routine; therefore, this information about tits as a species cannot be used to explain what this tit does.

This tit's behavioral goal is unique, so the judgment that this tit is acting in order to open the bottle top so as to drink the cream contains information regarding *this* tit. *It* is currently so constituted that there is something about it, we know not what, that motivates it to act with the proximate goal of opening the milk bottle so as to drink the cream. *That* this tit's behavior has this goal can be discerned through knowledge of the behavior and knowledge of the goals and resources available to tits. But *why* this tit acts with this goal is determined by the unique fact about this unique tit that it is motivated to open milk tops. That is, this tit *wants* to open this milk top.

But why does this tit act in just this way (pecking) when it attempts to act on its desire to open the milk top? The attribution of a desire to an agent is insufficient to account for what that agent does, of course. One can know that our tit is acting in order to open the bottle top because it wants the top to be open and still not know why she pecks at the top. To explain this, one also needs to know a further fact about this particular tit. What the tit does in attempting to open the bottle—peck at it—*would* open the bottle, were pecking capable of breaking through the foil. So there must be something about this tit, we know not what, that motivates it to act as if pecking on the top would break it open. That is, this tit *believes* that pecking will open the top.

Rational explanations are two-factor explanations. When one appeals to an agent's reasons to explain what the agent does, one mentions two different facts about that agent: its desire and its belief. But rational explanations differ from other two-factor explanations insofar as these two factors together provide a *justification* for the action as well as an explanation.

As we saw in the last chapter, the very notion of instrumental rationality demands that, in general, what an instrumentally rational agent does for reasons should attain the agent's ultimate ends. But what an agent should do to attain an end in a given situation is whatever has the effect of achieving that end in that situation. Because rational agents often act rationally in situations in which the available instrumentalities do not provide a direct route to their ultimate ends, an act would be justified in those situations if it resulted in some state of affairs that could be altered by a further act in such a way that it resulted in a state of affairs that could be altered in such a way that . . . the agent achieved its ultimate end. Each of the acts of an instrumentally rational agent is justified, insofar as it is justified, by its relations with other actual and potential rational—that is, justified—acts of the agent. So to say that an agent acts rationally is also to say that most of its acts are justified.

Most of what rational agents do is rational (justified), given the agent's ultimate ends. The two factors appealed to in a rational explanation, the agent's belief and desire that account for the act, explain the act by showing how the act is justified. The agent acts in this way because it would achieve the end specified in the desire, were the conditions specified in the belief actual; and, since most of what a rational agent believes is true, most of the time what the agent does is justified. The tit pecks at the top because it wants to open the top and believes, truly, that if it pecks on it, the top will be opened. Implicit in this explanation is the assumption that, in general, what the rational agent does is justified. The rational agent will perform mostly justified acts. This act is justified because it would lead to some end of the agent by such-and-such route if conditions were such-and-such and because those conditions are, in fact, actual. Since rational agents for the most part, do what is justified, by showing the way in which this act is justified one also explains why it is that the agent acts in this way. The tit believes truly that it can get through the top by pecking, it wants to open the top, and because the tit is rational, this is sufficient to explain why the tit pecks at the top.

As we saw in the last chapter, the contents of the beliefs and desires that provide the reasons for an agent's acts are relevant to the explanation of those acts insofar as they justify the acts. But not everything that a rational agent does for reasons is justified in the sense that it leads to the attainment of the agent's goals, ultimate or proximate. Rational agents make mistakes. Nevertheless, even the mistakes of a rational agent are justified.

When a rational agent makes a mistake, it does something that is not successful or that, if it is successful, is not successful for the agent's reasons. So the agent's reasons do not explain the success of the act. Nevertheless, the rational agent's act *is* goal-directed and thus has the interesting property of being such that it *would* have been successful at reaching some goal, had circumstances been different. And, we can ask, why did the agent do something, physically described, that had just this interesting property? The answer, of course, is that the act was brought about by just this belief and this desire and the act was justified *by* these reasons, even though the act itself wasn't justified. An act is "justified by" a belief/desire pair that explains that act if the act itself would be successful—would be justified—were what is believed by the agent actually to obtain. Notice that an *act* need not be justified simply because it is justified by the beliefs and desires that explain it. An act is justified if it really accomplishes its end; it is justified by its reasons if it would attain the object of desire *were* what is taken for granted in the act really to obtain, that is, were the belief "true." There is nothing more realistic and objective than the sort of truth required for a pragmatic theory of content, because there is nothing more realistic and objective than the success or failure of the acts of agents. Just ask any dead plover.

An explanatory factor counts as a reason for an act just in case it justifies that act. Reasons justify an act to the extent that they explain why the act occurs and to the extent that the act is also justified by those reasons. The two reasons appealed to in a rational explanation jointly justify the act insofar as one of the factors accounts for the act's having a certain goal and the other specifies the conditions under which the act would be successful. This pair of factors justifies the act in the sense that they explain why acting in just this way *could be considered* instrumental to reaching this goal. Different pairs of reasons justify different acts with different teleological roles, and, given the nature of justification, the factors that can serve as reasons for a given act vary as a function of the teleological role of the act. Jumping up and down can't be justified by the same pair of reasons by which walking to the store could be justified because they have different effects in the same possible situations. So when a rational agent acts rationally, its act is explained by the two facts about the agent that motivate it to act in a way that has a certain goal that would be obtained, were some particular definite conditions actual.

Rational agents act as they do *because of* their reasons. By itself, no belief or desire can explain why the agent acts as it does. Reasons explain only insofar as they justify, and it is because those reasons justify that they explain. The tit pecks at the bottle precisely *because* such pecking would open the top, were the tit's instrumental belief true. Because the tit is rational, it will act in a way that is justified by its reasons, and most of the acts that are justified by its reasons are themselves instrumental to attaining the tit's ultimate ends. The fact that the action is justified by agent's reasons explains why the rational tit pecks at the bottle top.

We explain rational action by offering a rational explanation of what the agent does. Agents that act in a rational way act as they do because they have reasons of their own for acting as they do. Those reasons are the unique characteristics of the individual agent that account for why the rational agent acts in order to realize some definite proximate goal by doing something in particular. The factor that motivates the rational agent to act with some definite proximate goal can be identified as precisely that: the factor that motivates acting with that goal, or the desire that that goal be achieved. The factor that motivates the rational agent to act in some determinate fashion in attempting to achieve its goals can be identified as the factor that motivates that agent to act in its attempt to achieve its goals as if some possible state of affairs were actual or as if the belief that that state of affairs actually obtained. That these factors, taken together, justify the act explains why a rational agent acts in that way.

The rational acts of rational agents are explained by alluding to their beliefs and desires that justify their actions. That is, those acts are explained by appealing to the agent's reasons for acting. Nonrational agents lack such distinctive, justifying motivational features, so what such agents do is not explicable by appeal to reasons of the agent's own. There are good reasons for what the wasp does, in the sense that certain actions of the wasp are likely to lead to the achievement of the wasp's intrinsic goals. But these reasons are not reasons that the wasp *has:* they are not reasons of the wasp's own. For the wasp does not act as she does because of her own beliefs and desires that justify her acts.

## 2. INFERENCE AND EXPLANATION

Every individual instrumentally rational act of an instrumentally rational agent must be performed because of the agent's reasons. That is, every such act takes place because of some belief and some desire of the agent that justify that act. This is part of what it means to say that the act is rational, or done for reasons. But this is not the only constraint on the rationality of action. The fact that instrumentally rational action is directed toward some goal implies that every individual rational action must fit together with others so as to form a coherent pattern of goal-directed behavior. The holistic character of what it is for an act to have a goal must be reflected in the nature of rational action, and this holistic character of rational action places important constraints on just which reasons a rational agent can have and which reasons can motivate rational behavior.

No act of any agent, rational or not, can have a goal in isolation. Even such a simple and straightforward action as the tit's pecking at the bottle has a goal only in the context of the rest of the tit's behavior and in relation to the fact that a tit is doing the pecking. That goals are holistic relational properties of acts is reflected in the impossibility of determining the reasons for an agent's acts in isolation.

Every rational act must be justified by its agent's reasons, but this constraint wildly underdetermines those reasons. While it is true that the tit's pecking can't be rationally explained by the tit's desire to buy bread and its belief that it can get bread by jumping up and down, this pecking, taken in isolation, could be the result of the tit's desire to buy bread and its false belief that it can buy bread by pecking at the bottle top. Since acts can be unsuccessful and beliefs can be false, a single act by itself can never fix the desire that rationalizes it and thus cannot fix the goal of the action.

But no action of the tit, or any rational agent, is ever done in isolation. The tit continues to peck at the top until the top is broken, and then it stops pecking. When and if the tit breaks through, it proceeds to insert its beak into the bottle and to drink the cream. This result, we know, brings about the tit's being nourished, and nourishment is one of the tit's intrinsic ends. So, we conclude, the entire exercise was for the sake of the tit's being nourished. The pecking must be for the sake of opening the bottle so that the tit can drink the contents. The proximate goal of—and thus the desire that partially motivates—the pecking behavior is intelligible only in the context of the overall behavior of the tit and in light of the fact that a tit is the agent. But because this goal makes sense only in context, that this act has this goal implies that other acts of the tit also have their own proximate goals that make rational sense in relation to the goal of this act and that, in turn, serve to make rational sense of this act.

The holistic and nested character of attributions of goals to acts guarantees that it is always possible to raise and answer further questions concerning an agent's proximate goals and that these, in turn, carry implications regarding the agent's desires. But that the tit has these desires and that it acts in the definite, specific ways it does, implies that the tit also has certain beliefs—for example, that it can break the top by pecking and that if it drinks the contents it will be nourished. Each act has a goal only in terms of a pattern of acts with goals, and in rational agents this pattern always includes a pattern of beliefs and desires.

The moral of this little explanatory story is that the attribution of desires and beliefs to rational agents is holistic because all goal-directed attributions to actions are holistic and that the desires and beliefs of rational agents are attributed as a way of explaining why rational agents act with the proximate goals they do.

But the beliefs and desires of rational agents that form the backdrop against which their individual acts can have proximate goals are not just *any* beliefs and desires. Because the agent is rational, these states are, in general, just those states that motivate rational action—that is, just those beliefs and desires that this agent should have, given its ultimate biologically determined ends, its resources, and the actual situation it confronts.

The acts, beliefs, and desires of rational agents thus form a holistic system, and this system has a fundamentally normative structure. Insofar as their behavior is rational, rational agents do what they should given their beliefs and desires. By and large, rational agents believe what they should believe, given their other beliefs and desires and the actual situation in the world. And, in general, rational agents desire what they should desire, given their other beliefs and desires and the kind of organism that they are.

Given that the rational agent is rational, that agent does what it does for reasons that justify its actions. But among the things that such agents do is to adopt new beliefs and desires that are appropriate to the changing circumstances of their worlds. If an agent's motivations did not change in response to changes in its environment, the agent could not be rational, because the rational agent, in general, acts in a way appropriate to its actual circumstances. Insofar as the agent is rational, those changes in motivation must occur for reasons of the agent's own. That is, the rational agent has the beliefs and desires that it should because *the agent has reasons* to have them. Insofar as an agent is rational, the fact that this agent has the beliefs and desires that it does is rationally explicable. The web of beliefs and desires that explain the actions of rational agents—and in terms of which any particular act of the agent has the goal it has—is itself a web that is spun for reasons. Regardless of whether there can be theoretical reason without practical reason, there can be no practical reason without theoretical reason. And this fact places important constraints on the types of complexes of beliefs and desires a rational agent can have and on the interpretation of the beliefs, desires, and proximate goals of the acts of rational agents.

Beliefs and desires motivate not only overt action but also the act of coming to have other beliefs and desires. As with any other rational explanation, motivation alone is not enough. Insofar as an organism is rational, it is motivated to come to have the beliefs and desires that it has reason to have. That is, the organism's coming to have a belief or desire is rationally explained by a belief/desire pair only if that pair justifies coming to have that belief.

But how can a belief/desire pair justify coming to have another belief or desire? As I have developed the concept, an overt act is justified by its motivating reasons just in case the act would be successful at reaching its goal—the object of the motivating desire—were its motivating belief true: that is, were the world actually the way that is taken for granted in the act. The notion of an act's being justified or not is thus tied to the teleology of the act. That is, if there is nothing at which the act aims, there is no possibility of the act's being justified by its reasons. Unless the act has a goal, there is no place for the notion of an act's being justified to grab hold. So to say that the act of a rational agent's coming to have a new belief or desire is rationally motivated and explicable implies that such an act has a goal. But what could the goal of such an act be?

There is a perfectly straightforward sense in which a rational agent's coming to have a new belief or desire—any new belief or desire—is goal-directed. Rational agents are organisms that get what they need by responding rapidly and appropriately to altering circumstances. As we have seen, this ability to keep acting in a successful (that is, justified) manner, regardless of the novelty of the situation, is explained by appeal to the agent's beliefs and desires. Only insofar as the agent's beliefs track the actual world and its desires remain appropriately articulated in the real circumstances, does the behavior of the agent remain successful and justified. This, then, is what coming to have a new belief or desire is *for.* The ultimate goal of any act of coming to have a new belief or desire is to remain attuned to the environment as it changes, that is, to remain *rational.* The ultimate criterion for the success of these acts is whether the behavior of the agent continues to be rational and successful, and the standard for such success is given by the ends of the agent, ends that are fixed by its organic nature.

The dictum that the point of coming to have new beliefs and desires is to remain rational is highly abstract. Fortunately, it is relatively easy to fill in some of the details. There is reason to form a new desire if one can achieve one of one's ends by first achieving the object of the new desire. Our genius tit has good reason to form the novel desire to open the bottle top if doing so is in fact instrumental to the tit's being able to nourish itself. There is reason for me to want to harvest mature wheat if I can use that wheat to make food. I have reason to form the desire to harvest the wheat if I want to make food and believe that I can do so if I harvest the wheat. If, in accordance with my belief, I make food from the harvested wheat, then my act of coming to want to harvest the wheat is justified by its instrumentality in keeping my behavior rational, that is, successful. If, contrary to my belief, I really *can't* make food from wheat, then my coming to want to harvest the wheat is justified by my reasons, even though the act of coming to have that desire is not itself justified.

Coming to have a new belief works in a slightly different manner. The role of beliefs in rational agency is to track the actual state of the world that is relevant to whether a given course of action will be instrumental in attaining the agent's desires. Because rational behavior is behavior that *is* successful in tracking actual circumstances so as to make it more likely that the agent will act appropriately to satisfy a desire and, ultimately, a need, and because an agent acts rationally in part because its beliefs do track the actual world, the point of coming to have a new belief is to accurately track the world. So, for example, consider the case in which a rational agent acts in a way that, given the actual current circumstances, does not achieve the agent's proximate or ultimate end. That is, consider the case in which the agent acts in a way that constitutes a mistake. Now, since rational agents act as they do because of the beliefs and desires that make up their reasons, a mistaken action is an action motivated by a false belief. Insofar as the agent is rational, it will respond to this failure as a mistake and will alter its beliefs to compensate for the failure. That is, rational agents respond to mistakes by altering their beliefs, and thus their actions, so as to track actual conditions and thereby to increase the likelihood of future success: rational agents learn from their errors. This is precisely the respect in which plovers differ from wasps and that gives us grounds for the view that plovers might be rational whereas wasps certainly are not. Plovers routinely engage in self-correcting behavior that gives us reason to believe that they respond to mistakes by changing their beliefs and, thus, their behavior. As Wooldridge has taught us, wasps never do that. For example, when a plover makes herself prominent to a predator and goes into the broken wing display, she expects—believes—that the predator will follow her. But if the predator does not follow her, her act has failed as a result of a mistaken belief. Insofar as the plover is rational, however, she recognizes her failure as a failure and changes her behavior in a way that indicates that she has abandoned her previous belief and formed the new belief that the predator has not noticed her display. This new behavior provides evidence that the plover is responding to her mistakes as mistakes by forming new, correct beliefs.[2]

This interest in having beliefs that track actual conditions in the world extends beyond tracking only actual conditions currently relevant to the satisfaction of the agent's current desires, however. Because a rational agent must be able to respond appropriately to a changing environment, such agents have a built-in interest in tracking any environmental conditions that could be relevant to their goals, as those goals change in accordance with conditions and regardless of whether these environmental factors are currently relevant. That is, the uncertainty of the environment to which rational animals must respond provides a *pragmatic* reason for rational agents to desire to have true beliefs, that is, beliefs that track conditions that *might* be relevant to the success of future activity. It is valuable for our tit to have the true beliefs that it could nourish itself by drinking the contents of the milk bottle and that it could drink the contents if it opens the top, because one never knows when one might find a way to open the top.

Since part of a rational animal's style of life, its way of making a living, is to act rationally and since coming to have true beliefs about the world is instrumentally useful to the project of acting rationally insofar as true beliefs about the world may be useful in future actions taken in changing circumstances, every rational agent *ought* to desire to come to have true beliefs. Since the principle of charity necessitates that rational agents generally desire what they ought, rational agents generally desire to come to have true beliefs.

For any act to be justified, it must at least be motivated by a belief/desire pair. We have just seen that rational agents always ought to desire to have true beliefs and that they generally do have that desire. But what beliefs are involved in the justification of acts of attempting to come to have true beliefs? As we saw in the last chapter, rational believers have two different ways of acquiring new true beliefs, through sensation and through inference.

It is hard to imagine any rational animal that didn't have generally reliable equipment for sensing relevant environmental conditions. To count as even a teleological agent, an animal must have some means of differential response to varying environmental circumstances, and rational animals are animals. Because the acts of rational agents depend on the agent's beliefs, this implies that the causal results of sensation are likely to be correct beliefs. This fact by itself, however, provides no rational *justification* for those beliefs. The sensory beliefs are merely likely to be true. But, given that these sensory beliefs are both likely to be true and potentially relevant to future action, a rational agent *ought* to believe that they are true, unless it has reason to think otherwise—and rational agents in general believe what they ought. So rational agents, in general, believe in the truth of the metaprinciple that what their senses lead them to believe is likely to be true. Rational agents want to believe the truth, so they have a justification for coming to believe that what their senses lead them to believe is generally true.

Rational agents are very good at instrumental action, action that reaches a desired result through successive transformations of the environment. Since what rational agents do is a function of their beliefs and desires, for them to act successfully through such chains of transformation they must have true beliefs regarding currently counterfactual circumstances, that is, regarding what would happen were some act performed in a situation that does not exist but she believes could be made to exist through some action of her own. Such beliefs cannot be justified by the reasons we have just mentioned, because they are not derivable from sensation, concerning as they do nonactual conditions. Such beliefs can, however, be justified by other beliefs of the rational agent.

If a rational agent comes rationally to believe some truth regarding a counterfactual situation that could not be derived from sensation, some other motivating factor must explain and justify this belief. Now, in fact, possible states of affairs are linked in such a way that if one or another of them happens to occur, certain others are inevitable or likely to occur. This must be the case because rational agents *in fact* can act in an instrumentally appropriate chain of events that is justified by true counterfactual beliefs, and, since such beliefs cannot be derived from sensation, they could not have rationally come to have those beliefs unless they could derive true beliefs from other true beliefs. This, in turn, would be impossible unless the truth of one belief were in fact to guarantee or make likely the truth of another, that is, unless the actual occurrence of some possible state of affairs guaranteed or made likely the actual occurrence of some other possible state of affairs. Given these factual links among possible states of affairs, rational agents ought to come to believe that a state of affairs that is guaranteed or made likely by some other state of affairs is in fact actual, if they believe that the first state of affairs is actual. That is, rational agents ought to come to believe what can be correctly *inferred* from what they already believe. And, since rational agents are rational, they believe what they ought.

It would be irrational to think that all of one's natural tendencies toward belief formation were infallible (that is, it is rational to believe in the revisability of the laws of inference that one is inclined to accept), but since a rational agent is rational, it ought to believe that most of the beliefs it has a strong inclination to infer from its other beliefs are true. Since rational agents for the most part believe what they ought, they accept, and act on, correct rules of inference.

So, when a rational agent wants to come to know a truth and believes both that "if A is true, then B is true" and that A is in fact true, then the act of coming to believe that B is true is justified by her reasons. As with any other act, the fact that this inference is justified by reasons does not guarantee the success of the act, in this case that B is true. After all, A could be false, or "if A, then B" could be false. Nevertheless, unless coming to believe in accordance with this form led mostly to success, that is, to coming to have true beliefs, the agent could not count as rational. Thus it is a necessary condition on rationality that most of what the agent comes to believe through inference is in fact justified.

Rational agents want to believe the truth and believe that what they believe because of sensation or inference is mostly true. Together, these metarules serve in the rational justification and explanation of their coming to have the beliefs and desires that they do. Rational agents have reasons for wanting and believing what they do, and these reasons both justify and explain the agents' beliefs and desires.

## 3. GOALS AND DESIRES

The beliefs and desires of rational agents are not events. They are standing states of those agents. An event happens at a time, whereas states can and do persist over an extended interval of time. That I trip over a rug is an event. That I am clumsy and disposed to trip over things is a state. Beliefs and desires are more like my being clumsy than they are like my tripping.

Because the beliefs and desires of instrumentally rational agents must be related in an inferentially rational manner, beliefs and desires must be states rather than events. Since an instrumentally rational agent, in general, believes what it ought to believe given its other beliefs and the actual state of the world, at any given time that agent will have a multitude of beliefs. These include beliefs about what the agent will be capable of effecting in various counterfactual situations and about what would happen were various alterations brought about in the agent's environment. Through appeals to these sorts of beliefs, we both explain what the agent does that is indirectly effective in attaining her intrinsic goals and account for the agent's desires to attain novel proximate ends. We account for the tit's behavior in opening the bottle top, for example, by mentioning her belief that, were she to peck at the top, she would be able to open it, and we account for her desire to open the top by appealing to her belief that, were the top open, she would be able to nourish herself. Any instrumentally rational agent always has a multitude of such inferential beliefs, because, in general, such an agent believes most of what follows from its other beliefs. But because the agent can actually do only one thing at a time, the agent cannot be motivated to act on all of these beliefs simultaneously. So it must be possible for the agent to have a belief at a time at which it doesn't directly affect behavior. These beliefs persist through time, rather than occur at a given time; they are states of the agent rather than events that occur to or in the agent.

Similarly, the desires of an agent must be states of the agent rather than events. Instrumentally rational agents, in general, desire what they ought, given their ultimate intrinsic ends, their beliefs, and the actual state of the world. These desires fall into two categories. Some of the ends that rational animals desire to attain are prescribed by their animal natures, while some of a rational agent's desires arise in the course of the rational animal's attempt to rationally attain its intrinsic ends. So, for example, the tit's drinking the cream has the goal of providing nourishment that helps the tit to survive. At the same time, however, rational agents are rational; such agents do what they do for reasons of their own. When the tit drinks the cream, it has reasons for acting as it does. The tit *wants* to be nourished and *believes* that acting as it does will help to nourish it. Since, in general, rational agents desire what they should given their intrinsic ends, rational agents, in general, always desire to attain those ends. But this implies that such desires are states of the agent, rather than events that occur at a time.

Because instrumentally rational agents always desire to achieve the biological ends intrinsic to their animal character, these desires are available to be used in the explanation and justification of the novel behaviors and desires that mark the agent as rational. Like simple goal-directed explanations, rational explanations tend to be nested. The tit pecks at the bottle in order to open the top so as to be able to drink the contents and thus to nourish itself. What the tit does when it pecks at the bottle top has the proximate goal of opening the bottle, the intermediate goal of eating, and the ultimate goal of nourishing itself. But the tit is rational, so what it does, it does for reasons of its own. Since the tit believes and desires what it ought, given its ultimate ends, its other beliefs, and the state of the world, the tit desires to eat and believes that if it eats certain kinds of food it will nourish itself; and since only by eating those foods can it attain the end of nourishing itself and it desires to open the bottle top because it believes that it can eat if it opens the top, and it desires to eat. The instrumentally rational agent's standing desire for a less proximate end at each stage helps to justify, and thus to explain rationally, the agent's coming to have the desire to achieve a more proximate, but instrumentally novel and valuable, end and thus helps to explain the agent's actual behavior. But since any desired end can become a less proximate end achieved by first doing something else, all desires must be seen as temporally persisting states of the rational agent rather than as events that occur in or to the agent.

Beliefs and desires are thus states of rational agents rather than events. As such, beliefs and desires are to be distinguished from the coming of an agent to have a belief or a desire—for these latter are surely events. When our first rational tit noticed the top of the milk bottle, the tit's sensations caused it to have beliefs it previously had lacked. When the tit hit on the plan for nourishing itself by opening the bottle top and drinking the contents, it came by inference to have a new set of beliefs and desires that it previously did not have. Both of these were events. But the beliefs and desires that the tit came to have were not events. They were states.

The fact that beliefs and desires are states of the rational agent helps to explain why most of a rational agent's acts have the goals that they do in virtue of the desires of the agent that explain those acts. What the *Sphex* wasp does has the goal that it has in virtue of its role in the stereotypical life drama of *Sphex* wasps. But most of the instrumentally rational behavior of a rational agent forms no part of the stereotypical life of biological organisms of the agent's type. When a tit pecks at a bottle top, the goal of that act is not fixed by the stereotypical role that acts of that kind play in the life history of tits, because acts of that kind play no such role. This is even more obvious when we consider the incredibly more indirect instrumental behavior of humans. When an engineer sits at a drawing board, designing a dam, the goal of that behavior is not a function of its fixed role in the way human animals make a living. The behavior occupies no such role. In both cases, the goal of the act depends on the desire of the agent that is relevant to its explanation. But given the position presented here that no agent can have reasons for what it does unless it first has intrinsic biological goals, one might wonder how this is possible.

The answer, of course, turns on the holistic and relatively stable character of the web of beliefs and desires that constitute a rational agent's potential reasons for acting. Instrumentally rational agents generally believe and desire what they ought, given their intrinsic biological goals. These beliefs and desires are persisting states of the individual. Because the agent is rational, these states will be altered for rational reasons as the actual conditions in the world change. But in the absence of such changes, because the agent is rational and because the agent for the most part believes and desires what it ought, most of the agent's beliefs and desires tend to persist. As persisting states, these beliefs and desires of rational agents needn't continually motivate actual behavior. Most of what a person—or a tit—believes or desires at a given time is not directly relevant to what that agent does at that time. Nevertheless, since the agent *has* those beliefs and desires continuously, once the agent comes to have a belief or desire, that state is *available* as a possible source of rational motivation for action. These standing states of the agent can thus be used by an observer to interpret and explain the otherwise puzzling teleological activity of the agent.

In essence, rational agents *themselves* generate novel beliefs and desires from the beliefs and desires that they have solely in virtue of their biological and sensory makeup. This ability imparts the further capacity to act in novel instrumental ways in attempting to achieve the ultimate ends. By generating new desires, they generate new goals for their actions, and by generating new beliefs, they generate new methods of attaining those goals. But there are strict constraints by which instrumentally rational agents must abide in generating these new beliefs, desires, goals, and methods. These are the norms of instrumental rationality itself, which themselves prescribe norms of inferential rationality. Because instrumentally rational agents, in general, abide by these norms, these agents in fact are motivated to act in an instrumentally rational fashion by the beliefs and desires that they come to have through abiding by these norms of rationality. Because they are in fact motivated by beliefs and desires that they come to have through the operation of instrumental rationality, what the agents do can have a goal and an instrumental role in the ongoing instrumental behavior of the agents in virtue of the desire that motivates the act rather than in virtue of a standard role that this kind of act plays in the life history of organisms of the agents' type. It is only because the instrumentally rational agent is an animal with intrinsic goals *qua* animal that what the agent does has any goal at all. For the desires of the rational agent that fix the goals of the agent's rational acts are themselves generated only by applying the norms of rationality to the stabile goals and beliefs that the agent has in virtue of being an animal of a certain kind.

The same facts about instrumentally rational agents explain how an observer can come to interpret, explain, and predict what the rational agent does and how the observer comes to understand the purposes that the behavior serves. Since the agent is rational, it will come to desire and believe what it should, given its intrinsic ends, and will, in general, come to act in accordance with those beliefs and desires. An observer can know both what sort of organic agent an agent is and that the agent is rational and thus can know that the agent generates new beliefs and desires rationally. Since the observer is rational, the observer can infer from these facts a set of beliefs and desires that the agent ought to have, given the actual state of the world. (Of course, since the observer is only incompletely rational, this ideal normative estimate regarding the agent's mental states will itself be only partially accurate.) The observer can then look at the agent's actual behavior and assign explanations to each of the agent's acts from the store of beliefs and desires that the agent ought to have.

As the agent continues to act and as the world continues to change, the observer can and must revise this ideal theory of the agent's motivational states. But, alas, no agent is entirely rational, and even instrumentally rational agents sometimes do things they ought not, given their ultimate ends. Since the agent remains rational, however, the observer can hope to assign a rational explanation even for these errors by assigning false beliefs and inappropriate desires to the agent and testing these hypotheses to determine whether such attribution is consistent with the agent's overall behavior.

The result of these modifications is what Davidson calls the observer's teleological theory of the rational agent's behavior. At any given time, a rational agent capable of having beliefs about the reasons of other rational agents has a theory that attributes a whole web of beliefs and desires to another rational agent, and this theory can be used to predict and control what those agents will do in various circumstances. By observing the behavior of intruders, a rational plover, for example, might come to believe that an intruder will follow her if the intruder believes that she is injured, because the intruder wants to eat her and believes that this will be possible if the plover is injured. If the plover is rational, she will continually update this teleological theory of the intruder on the basis of the ongoing behavior of the intruder.

But, as we have seen, no such theory is possible unless the agent has intrinsic ends of its own in virtue of its biological nature that anchor the entire process of belief/desire generation in the agent, and thereby allow for the possibility of the interpretation of the subject's reasons by an observer.

## 4. THE MOTIVATION OF RATIONAL BEHAVIOR

The fact that beliefs and desires are states rather than events generates a puzzle: how can such states explain rational behavior? As states, beliefs and desires are persistently present in an agent *over* time, but we appeal to these states to explain why the agent acts as it does *at* a time. But if the agent had the beliefs and desires that explain its act at time t even at times other than t, then why didn't the agent act in the same way at those other times? I solve this puzzle by developing a new account of the explanatory force of rational explanations.

This puzzle motivates an important tradition, one that has attained a dominant position in contemporary philosophy. This tradition understands rational explanation as a species of causal explanation. By resolving the puzzle in the way I do, I undercut the reasons behind this tradition. I conclude that rational explanations are not causal and that the reasons that rationally explain an agent's acts are not the causes of those acts.

The role of causality in rational explanation and what it is for a state of an agent to have content turn on three central facts. These facts broadly concern the respective semantics of assertions of causal relation and of the rational explanation of behavior. First, statements that assert that two events are causally related are extensional, whereas claims that advance rational explanations are intensional. Second, the referring expressions that appear in the explanans of a rational explanation do not refer to the sorts of entities that can enter into causal relations, events. Third, rational explanations explain the truth of action sentences, and action sentences are existential generalizations, not individual sentences. Taken together, these three facts severely limit the importance of causality to the structure of rational explanation.

Statements that assert that one event causes another are thoroughly extensional. How the events are described makes no difference to the truth-value of the assertion. If event A causes event B and if event A is also the event that was mentioned on page two of yesterday's *Times* and event B is also the event that is referred to on page three of today's *Times*, then the event mentioned on page two of yesterday's paper caused the event referred to on page three of today's paper. Rational explanations, on the other hand, are intensional. How the factors mentioned in the explanans and explanandum are described makes a great difference to the adequacy of the explanation. My statement that the event mentioned on page two of yesterday's paper caused the event referred to on page three of today's might be true and thus might provide a causal explanation, but as it stands it provides no rational explanation. One has provided a rational explanation only if one shows how the factors appealed to in the explanans justify the fact mentioned in the explanandum: in the example above, this does not happen. Of course, there might be other descriptions of events A and B that, if marshaled in the appropriate way, could be used in an adequate rational explanation of B by A. And it might be a necessary condition on the adequacy of such a rational explanation that A cause B (after all, I myself have suggested that it is a necessary condition on this sort of explanation that A *motivate* B). But given the intensional character of rational explanations, merely mentioning that A causes B can never be sufficient for giving a rational explanation of B.

But it is also true that A's causing B can never be *necessary* for A rationally to explain B: in the most typical cases of rational explanation of B by A, it isn't even *possible* for A to cause B. Simply put, the ontology of A is all wrong for such a causal connection. Causal relations are always relations between particular *events*. It might be the case, as many think, that one event can cause another only if both events belong to types that are related by some causal law. But it is always one particular event of a given type that causes another event of some other type. The factors invoked in rational explanations typically are *not* individual events, however. While it is true that we sometimes appeal to events in giving a rational explanation (for example, my noticing that the lion is in the grass could be part of a rational explanation of my getting back in the car), in the more typical cases rational explanations appeal to *states* in or of the agent. When I appeal to my beliefs and desires in explaining my going to the store, for example, neither my belief that if I go to the store I can get bread nor my desire to get bread are, as such, events at all. Rather, they are standing states of myself, the agent, and, as such, they don't enter into causal relations with anything.

Rational explanations do not assert that the agent's reasons cause the agent's actions. Reasons are not events and thus *can't* cause actions. Note that this does not depend on the semantics of the way in which rational explanations are presented. No matter how one describes reasons, they are what they are: states. So reasons aren't causes.[3]

But this result leads to a problem, and this problem historically has led to the received view that reasons are at least closely associated with causes. Beliefs and desires are states of a rational agent that explain the behavior and the rationality of the behavior of that agent. As discussed above, agents have reasons for what they do on one occasion only if they have reasons for what they do on many occasions, and only if those reasons are related according to the norms of inferential rationality. Given this holistically inferential character of reasons, rational agents must have, at any time, beliefs and desires that they are not currently acting on. That is, there is a difference between having a reason and acting because of a reason. That an agent has a reason can be gleaned from an overall interpretation of a substantial slice of the agent's rational behavior, but one cannot infer from the fact that an agent has a belief/desire pair that would justify acting in some way that the agent will be motivated to act in that way by that pair of reasons. This implies that what it is for an agent to have a reason can't require that the reason in fact cause any behavior. But it also implies that something more than the agent's having a reason to act is involved when the agent in fact does act on the reason. When one offers a rational explanation, the argument proceeds, one is explaining that some act, some event, occurs. And the only way to explain why an event occurs is causally. So rational explanations of an action must implicitly invoke some event that causes the action. Since this event is a necessary feature of the explanation, there must be some relation between the belief and desire mentioned in the explanation and the event that causes the act being explained. That is, the difference between having a belief/desire pair and acting on it is that, in the second case, the reason is "associated" with some event that actually causes the action.[4]

The character of this "association" is something of a mystery. The mystery arises because rational explanations necessarily involve both justification and motivation. Since what is being explained is the occurrence of some event—the act—it appears that the explanation must, in some respect, be causal and that because it is, rational explanations must appeal to events. At the same time, however, rational explanations are explanatory only to the extent that they show why the act being explained is something that the agent *should* do. We understand why this event happened when we see that it's occurrence would lead to a certain result under certain conditions, that the agent takes those conditions to be actual, and that the agent is attempting to reach that result. These factors in the agent are not events, but standing states, and by themselves these states do not motivate the actions they explain.

So it seems that the factors appealed to in rational explanations must both be, and not be, events. In the simplest model, when an agent acts on a reason, the reason is an event that itself is the cause of the act. This is classical token identity. As we have seen, however, this view is precluded by the fact that reasons as reasons are not events. At best, when they motivate action, they are "associated" with the events that cause the action.

Oddly enough, how we understand this association doesn't make much difference to a proper understanding of rational explanation because of the logical character of the explanandum of rational explanations. The logic here is tricky, and one must be careful. Explanations explain why some proposition is true. Rational explanations explain why some action sentence is true. It is natural to assume, as the received view does, that action sentences are individual sentences about particular nameable events. The assumption is that the sentence "I went to the store yesterday" asserts that some one particular event—my actual trip to the store, in all its particularity—happened yesterday. This particular event happened at a particular time, of course, so we assume that to explain why it is true that I went to the store yesterday one must explain the occurrence of this particular event. But this assumption is false. It is false because, appearances to the contrary, the sentence "I went to the store yesterday" is an existential generalization.[5] It really asserts something like, "There is something such that it occurred yesterday and was a going to the store by me." To explain why this is true, one need only show why, at some point yesterday, I went to the store.

It is, of course, possible that the sentence "There is something such that it occurred yesterday and was a going to the store by me" is made true by a single, unique event. This will be the case whenever I go to the store only a single time on the day in question. But since all that is necessary to *explain* why I went to the store yesterday is the ability to infer that the statement "There is something such that it occurred yesterday and was a going to the store by me" is true, in order to explain why I went to the store yesterday one need not explain why that very event, individuated by its causes and effects, occurred. Rather, all that is required is to be able to show why there is *some* event that satisfies that description.

The description involved in the existential generalization that I went to the store yesterday does contain a temporal parameter, and this is fairly typical. But the parameter is quite broad, and this is also quite typical. Since the generalization is made true by *any* event that occurred yesterday and that satisfies the rest of the description, as long as one explains why *some* such event occurred yesterday, one has offered all the explanation necessary. To be adequate, the explanation need not explain why just the very event that did occur occurred. And insofar as reasons *themselves* often contain a temporal dimension, rational explanations can be entirely adequate even independent of any reference to a particular event that triggers, or causes, the single event that satisfies the description of a going to the store by me yesterday. For example, in the case before us, we can add the further rational considerations that I came to have the perceptual belief yesterday that I was out of bread and that I have the standing desire that there be bread in the house because I have the standing belief that if there is bread in the house I can make my children's lunch (and can't do this if there isn't any bread). These further reasons, together with the fact that as a rational agent I usually act as I have reason to, help to explain why at *some* point *yesterday* I went to the store. The further fact that at the time I went to the store there was some additional event that triggered the particular event that satisfies the action description is irrelevant to the *rational* explanation of the truth of the action sentence. That there is *some* such triggering event is guaranteed by the fact that as a rational agent I act on reasons, and I had reasons to go to the store yesterday. But which event that is and when it occurred and what in particular caused it are all irrelevant to the rational explanation of the truth of the action sentence. And since it is irrelevant to *that* explanation, it is irrelevant to the content of the attitudes that rationally explain the action.

Because rational explanations explain why some existential generalization is true, they do not explain why some *particular* event occurs. To the extent that the action explained, in order to be the action that it is, must occur during some interval of time (I went to the store yesterday), the rational explanation of that action must appeal to reasons that have a temporal dimension ("I believed yesterday that P" or "I believe that P is true at time t" or "I desired D yesterday" or "I want that D should be true tomorrow"); but no appeal to reasons ever explains why just a particular event, individuated by its causes and effects and designated by a rigid designator, occurs. Because this is so, rational explanations as such are not causal. It is true, of course, that every time a rational explanation explains an action—that is, points out why some statement that there is an event of a certain type is true—there is some particular event that makes that generalization true. And there is surely some causal explanation for why just that event occurs. But this explanation is not part of the rational explanation. The rational explanation only displays why a creature of this organic type should, under the circumstances, do *something* that has this description and appeals to the further fact that this creature is rational—that is, that it mostly does what it should—to reach the conclusion that this agent will do something that has this description.

When an agent acts on a reason, it acts in accordance with the norms of practical rationality, and what we are saying when we say that the agent has beliefs and desires is exhausted by this fact. The agent has just the beliefs and desires it must have if its action is to be rational in just the way that it is. When the agent performs a particular rational act, this act is *motivated* by just those beliefs and desires that rationalize it. This is what it is for an act to be rationally motivated by reasons. At the same time, this rationally motivated act is *triggered* by some event in the sense that it is caused by that event. But just how this triggering event is "associated" with the motivating reasons is irrelevant to those reasons. The triggering event explains why just this event occurs. The motivating reasons explain why an event of a certain type occurs. And while the second explanation implies that there is *some* triggering event, just which event that is or how that event is to be described (over and above "the event that triggers an action in accordance with these reasons") makes no difference.

This looseness between rational explanations and the triggering causal explanations that they imply explains a curious but well-noticed fact about the ontology of beliefs and desires. It is tempting to treat attitudes as *dispositional* states. After all, if someone believes that P, she will in general be disposed to act as if P, and if she desires D, she will in general act in ways that result in D. But when one tries to specify exactly what disposition a belief or desire is, it quickly becomes apparent that such specification is impossible. Any ordinary disposition is determined as a set of ordered pairs in which the first member of the pair is a possible set of affairs and the second is a result that will be brought about in that set of circumstances. So if it is true that some entity is in a dispositional state S, which is partially defined by the pair C-R, and that circumstances C obtain, then it is entailed that R will result. But it is never the case that if a rational agent has some belief B, there are some circumstances in which it must do some R. The logical behaviorists discovered to their chagrin that this is true when C and R are specified in physical terms. Only belief and desire together rationally explain action. But this is equally the case even when C and R are allowed to be intentional descriptions. After all, one can have reason to act and not act, either because one has a stronger reason for not acting or for no reason at all. And since the stronger reasons for not acting are infinitely various—as are the possible causes of not acting—there is no finite set of ceteris paribus conditions that can be built in to the definition of the dispositional property that was meant to specify the attitude.

But we can see now why this is the case and why it is not quite true to say that beliefs and desires are dispositions. When one offers a rational explanation of an act, one is appealing to the fact that in general the agent can be expected to act in ways that have certain descriptions because the agent should act in ways that have those descriptions. These are the various ways of acting that are consistent with and implied by the agent's real situation, the ways the agent has acted in the past, and the agent's organically fixed ends. After the agent acts, we explain the act by noting that it is one of the things that fits one of these descriptions. And before the agent acts, we can predict that it will do certain things within certain temporal limits. But unless the agent finds itself in extreme circumstances, at any given time there are quite a variety of things that it *could* do that, under the circumstances, would be rational to do. After all, at this point (it is after 3:00 in the afternoon), either my calling it a day and going to the store to get that bread I want or my finishing this section might be rational. If I go to the store, I would have good reasons for doing so that explain *that* act and that, in addition, happened to motivate that particular act at this time. But if I stay here and finish the section, I would also have good reasons for doing so that, as a matter of fact, motivated *that* act. My beliefs and desires regarding these subjects are pretty vague in their temporal dimensions ("I need the bread today and I want to be home by 5:00. It takes an hour to get home, hum . . . but it would be good to finish this today even though I know pretty much what I am going to say and I *could* do it tomorrow") and, by themselves, don't determine what, in particular, I will do now. They don't determine this because either act is consistent with my rationality.

*What* in particular I do will be triggered by some event, but rational explanation knows nothing about *that event*. It might be motivated by states with content, or it might not. In any case, I can have the beliefs and desires in question regardless of whether I am triggered to act on them. But even though these triggers are irrelevant to whether I have the beliefs and desires that I do, they are central to whether I have the dispositions I do. For those dispositions are just *defined* by what I will do in given circumstances, and these triggers, as the causes of my action, are central to the definition of those circumstances.

So we can conclude that beliefs and desires are not *quite* dispositional states, although they certainly share features in common with dispositions. They are states of an agent that would be relevant to the rational explanation of the agent's action, were the agent motivated to act in a way that is justified by the content of those states.

## 5. RATIONAL EXPLANATION AND INTENTIONAL CONTENT

Beliefs and desires are states of an instrumentally rational agent that would be relevant to the rational explanation of the agent's actions, were the agent motivated to act in a way that is justified by the content of those states. At any time, every instrumentally rational agent is characterized by the possession of a complex set of motivationally relevant states that are related to one another and to the agent's behavior and organically fixed goals in exceedingly complex ways. Because what the instrumentally rational agent does can be interpreted, predicted, and explained if an observer understands these states and these relations, there are excellent instrumental reasons for any instrumentally rational agent to come to have true beliefs about these states and these relations.

We human, linguistic agents have developed a sensitive system for keeping track of these states and relations. We say that these states fall into two main categories: beliefs and desires. A desire is a state of an agent that is relevant to the explanation of the fact that an instrumentally rational agent acts with some goal. A belief is a state of an agent that is relevant to the explanation of the fact that an instrumentally rational agent acts in attaining the ends of its desire in the particular ways that it does. Because instrumentally rational agents engage in many different acts with many different proximate goals, instrumentally rational agents have many different states that help to account for this diversity: they have many different desires. Because instrumentally rational agents act as if many different states of affairs were actual, instrumentally rational agents have many different states that help to account for this diversity: they have many different beliefs.

What differentiates one desire from another is the relevance of one desire to the explanation of the fact that the agent acts with one goal and the relevance of another desire to the explanation of the fact that the agent acts with another. That is, desires are individuated by their *contents*, by the fact that one desire can motivate and help to explain an action that has the goal of achieving p and another desire can motivate and help to explain an action that has the goal of achieving q. What differentiates one belief from another is the relevance of one belief to the explanation of the fact that the agent acts in detail to attain its desires as if one state of affairs obtained and the relevance of another belief to the explanation of the fact the agent acts as if another state of affairs obtained. That is, beliefs are individuated by their *contents*, by the fact that one belief can motivate and help to explain an action that presupposes that p is true and another belief can motivate and help to explain an action that presupposes that q is true.

The beliefs and desires of an instrumentally rational agent are relevant to the prediction and explanation of the agent's behavior only insofar as they are related to one another, to the environment, and to the acts of the agent in certain normatively complex ways. We keep track of those relations by recognizing that an instrumentally rational agent's beliefs and desires, as well as its rational activity, obey the constraints of the principles of charity. Therefore we are capable of recognizing that instrumentally rational agents frequently have the desires that they ought to have, even though those desires are not prescribed by the agents' biology. This in turn allows us to understand when an agent is acting for a goal that is not prescribed by its biology, which in turn allows for the possibility of recognizing that the agent has yet other desires and beliefs. So the system of bookkeeping through which we keep track of the motivationally important states of rational agents by assigning contents to those states can become incredibly complex.

In particular, the requirement that the system of an instrumentally rational agent's beliefs and desires obey the constraints of the norms of inferential rationality adds a layer of considerable complexity. When an agent believes that p, it is rationally motivated to act as if the world is actually p. But, given that instrumentally rational agents desire to come to have true beliefs, a significant part of what it is for a rational agent to act as if p is the case is for that agent to be rationally motivated to come to believe that whatever follows from p is also the case. Thus, since for an agent to believe that p is for that agent to act for its goals as if p were the case and for an instrumentally rational agent to act for goals as if p were the case is to a large extent to act in such a way as to infer what follows from p, to a very large extent what it is for an agent to have a belief with the content "that p" is for an agent to be in a state that motivates correct inferences from p. And since q can follow from p—or from p and r, or from p only if s, or . . . —what an agent who believes p has reason to come to believe in virtue of that fact, and in what circumstances, overt and psychological, it has that reason to come to have those beliefs, is exceedingly complex. So the content of the beliefs of instrumentally rational agents is to a very large extent fixed by the inferential role of those beliefs, and any specification of those roles will be extremely complex. Nevertheless, in every case, when one says that an agent believes that p or has a belief with the content "that p" one is saying that the agent will, in general, act as if p is the case, including acting so as to come to have new beliefs that can correctly be inferred from p's being the case. And when one says that some agent has the desire that p, or has the desire with the content "that p," one is saying that the agent will, in general, be motivated to act with the goal of bringing about p.

## 6. SUMMARY

At this point, it is thus appropriate to say that we have developed the outlines of a new, pragmatic account of what it is for an agent to have beliefs and desires that have intentional content. This new account offers an alternative to both standard causal-role functionalism and the sort of representationalism championed in different ways by Jerry Fodor, Fred Dretske, and Ruth Millikan. In contrast to representationalism, this pragmatic account asserts that the content of an intentional state of an agent arises out of the role of that state in the explanation of and rational justification for the actions of the agent, rather than being a feature a state can have independent from its role in the explanation of action. In contrast to causal-role functionalism, this new account posits that the relations among the states of the agent, the agent's acts, and the environment that determine the contents of states are primarily normative rather than causal and that the acts explained by these states admit of teleological as well as physical descriptions.

In chapter 1, I specified three mysterious aspects of intentionality that any account of intentionality must confront. How does intentionality fit into the physical world? Is believing that p or desiring that q an intrinsic or relational property of agents? How is it possible to understand the intrinsic normativity associated with intentional content? This pragmatic account offers answers to each of these questions.

In the pragmatic account of intentionality, for some state of an agent to be a belief or desire with intentional content is for that state to occupy a certain role in a pattern of relations among the behavior of an agent and the states of that agent that explains the behavior. In this respect, the pragmatic account is similar to more familiar functionalist accounts of intentionality. The pragmatic account is distinctive in that the pattern of relations among these factors is understood to be one that satisfies a variety of normative constraints rather than one consisting in a set of causal roles. Nevertheless, the pragmatic account shares with functionalist alternatives the view that a state's being a belief or desire with intentional content is essentially a matter of that state's having an abstract relational property. Because being a belief or a desire with content is a matter of a state's having an abstract relational property, there is no ontological reason that some physical state could not also be an intentional state. In addition, in light of the fact that the pragmatic account does not see rational explanation as asserting a causal connection between an agent's reasons for acting and the agent's acts, it stands under no obligation to supply a solution to the so-called problem of mental causation. Appeals to beliefs and desires *explain* rational behavior; but beliefs and desires don't cause that behavior.

The pragmatic account also makes the apparently relational character of intentionality comprehensible. Just as an act can have the goal of achieving p without that act's involving a real relation to any actual or possible p, a state can be the belief that p or the desire for p without that state's involving a relation to any real p or any representation of p. When one says that some act has the goal of achieving p, one is saying that the act occupies a certain role in an ongoing pattern of relations among the acts of an organic agent of a certain sort. The reference to p in the statement is a way of keeping track of and identifying just which role that act occupies in that pattern, but it doesn't imply that the act is related to p. So an act's having a goal is indeed a relational property of that act. But that relation isn't a relation between the act and *p*. Rather, it is a complex holistic relation among the act, other acts of the agent, the agent itself, and the agent's environment. Similarly, when one says that some agent believes that p or desires q, one is saying that there is some state in that agent that occupies a certain role in an ongoing pattern of relations among the instrumentally rational acts of the agent and the states of the agent that rationally explain that rational behavior. The reference to p in the statement is a way of keeping track of and identifying just which role that state occupies in that pattern, but it doesn't imply that the state is related to p. So a state's having a content is indeed a relational property of that state. But that relation isn't a relation between the state and *p*. Rather, it is a complex holistic relation among the state, other states of the agent, the rational behavior of the agent, the agent itself, and the agent's environment.

Finally, the pragmatic account of intentionality renders the normative character of intentionality understandable. Just as the characterization of an act as having a goal ipso facto renders it capable of being evaluated normatively in terms of its success and failure at reaching that goal, so the characterization of a state of an agent in terms of its intentional content also renders it capable of being evaluated normatively in terms of its content. The reason for this is the same in both cases. Both teleological and intentional descriptions are inherently *normative* descriptions. To say that an act has a goal or that a state has a content is to say that it belongs to a pattern of relations that mostly satisfies normative constraints but that has elements that deviate from those norms. Because to be a goal-directed act or to be a belief or desire is to belong to such a potentially broken pattern, it is always possible to ask—and answer—the question whether this act, or this state, satisfies the norms constitutive of the pattern or whether it deviates from them. That is, it is always possible to normatively evaluate the act or the state.

It is the fact that the pragmatic account of intentionality can help us to understand the normative dimension of intentionality better than any of its rivals that above all gives us reason to believe that it is true.

CHAPTER SEVEN

# PRACTICAL REASON AND DEFINITE CONTENT

## 1. THE LIMITATIONS OF MERELY INSTRUMENTAL REASON

The pragmatic account of intentional content has several distinctive features. First, and perhaps most important, this account makes good, nonmetaphorical sense of the notion that organisms act in order to accomplish goals, even when those organisms are incapable of rational action and have no beliefs and desires with intentional content. What it is for an act to have a goal is a complex, holistic property of the act, one that involves relations among the act itself, other acts of the agent, the fixed organic ends of the agent (for example, survival and reproduction), and the typical life pattern of this kind (species) of agent. Furthermore, on this pragmatic account, instrumentally rational action is a species of goal-directed behavior that is flexible, versatile, and appropriate in novel ways to achieving the fixed organic ends of agents. Because instrumentally rational action involves novel responses to changing environments, what it is for any such rational act to have the goal it has cannot be fixed relative to the stereotypical pattern of behavior of the agent's species. Instead, instrumentally rational action involves behavioral epicycles in which the rational agent achieves some ultimate fixed end by first acting to achieve some novel proximate end(s). Application of the principle of charity, together with an understanding of the available means and ultimate ends of the agent, allows for the possibility of fixing the proximate goals of some instrumentally rational acts.

On this view, beliefs and desires are the reasons that explain and justify the acts of rational agents. In the simplest cases, a rational agent desires to achieve the proximate goals of its acts and believes that the world is in the possible state that, if it were actual, the agent's act would bring about a state of affairs that would satisfy its desire. The requirement that most of an instrumentally rational agent's acts be successful implies a series of normative constraints on how beliefs and desires must be related to each other, the acts of the agent, and the agent's environment. These constraints allow an observer who knows the beliefs and desires that motivate some of the instrumentally rational activity of an agent to infer, in bootstrap fashion, the rational motivations for other acts of the agent. This inference, in turn, allows holistic specification of the contents of an agent's beliefs and desires and the possibility for the understanding, explanation, prediction, and justification of acts of the instrumentally rational agent that would otherwise be unintelligible. The contents of a rational agent's beliefs and desires can thus be seen as arising out of the roles of those states in the rational explanation of the overall instrumentally rational behavior of the agent.

Although this account as presently developed has many virtues, it is inadequate as an account of human rationality and of the kinds of contents typical of the intentional states of human agents. This inadequacy can be articulated in the form of two objections.

First, the kinds of contents that can be attributed to the beliefs and desires of instrumentally rational agents on the grounds that I have articulated are hopelessly vague when compared to the kinds of contents we routinely attribute to human rational agents. While a tit, I would argue, may have the desire to drink the contents of some milk bottle, there is simply no sense in asking whether what it wants to drink is the cream or the potable liquid or the nourishing liquid or what have you. There is no sense in raising this issue because, if the account I have offered is on the right track, the grounds on which we attribute desires to tits offer us no possibility for making such distinctions. While the contents of a tit's beliefs and desires might be opaque to *some degree,* they are surely not *that* opaque. But any account of content that is inadequate to understand the level of opacity routinely found in human intentional states is inadequate overall.

There is a second obvious problem with this account. The pragmatic aspect of the account ties rationality inextricably to *instrumental* rationality. Beyond this, the account tells a crucially naturalistic story regarding instrumental rationality itself, a story that sees the ends for which instrumentally rational agents act as rooted in the organic interests of the agent as an organism. It could be argued (I think correctly) that such a view of reason is doubly wrong regarding human rationality. Whatever is true of other animals, human beings, at least, sometimes act in order to accomplish goals that not only are not rooted in their interests as organisms but actually are inimical to those interests. Ascetics who take a vow of chastity or zealots who welcome martyrdom for some cause cannot be seen as doing what they do in order to reproduce or survive, no matter how indirectly one traces out the connections. It also seems that, at least on occasion, human beings act rationally in a way that is not instrumental at all. Arguably, when the ascetic sacrifices her reproductive capacity in order to become a nun, she does so not in order to realize *any* concrete state of affairs, no matter how described. Instead, she does so for the sake of being a certain sort of *person*. This sort of reason is not properly understood as an instrumental reason at all. Now, it would seem at first glance as if the kind of naturalistic, instrumental account of rationality that I have developed cannot make sense of even the possibility of ends that are not rooted in the nature of an organism or of a type of rationality that is not instrumental. But if this is so, and noninstrumental and nonorganically based ends are actual in human beings, then it would seem that this account must be wrong.

## 2. OPACITY AND VAGUE CONTENTS

The standard criterion for attributing intentional content is intensionality. Leaving aside modal contexts, if the attribution of a state to an object involves a context in which the description of an entity or state of affairs makes a difference to the truth of the attribution, then that attribution involves the ascription of content to that state. Ordinarily, the description of an entity makes no difference to a determination whether some statement including that description is true; the description is transparent in the sense that what renders the statement true or false is the entity or state of affairs described, no matter *how* it is described. If Valerie is taller than Anna, and if Valerie is my oldest daughter and Anna is my youngest daughter, then my oldest daughter is taller than my youngest daughter. But it can be true that you can believe that Valerie is taller than Anna and not true that you believe that my oldest daughter is taller than my youngest daughter, if, for example, you think (falsely) that Kathryn, and not Anna, is my youngest daughter. So your belief is intentional: it has the intentional content that Valerie is taller than Anna but does not have the content that my oldest daughter is taller than my youngest. The fact that what we are speaking about is your *belief* removes the transparency of the descriptions of my daughters and renders them opaque.

At which point in the evolution of organic nature does opacity, and thus intentional content, begin? In one sense, this question is ambiguous. On the one hand, when one raises this question, one might be trying to decide whether it is possible to discern in our attributions to other animals the level of intensionality and thus crispness of content that we routinely find in our attributions of thoughts to human beings. On the other hand, when one raises this question, one might be asking about the point at which our attributions to other animals display *any* opacity at all. These are two very different questions with two very different answers. The answer to the first question, the one regarding the relative determinateness of the content of human and animal thoughts, leads to the first objection to the view of content that I have mentioned. The answer to the second question, concerning how deeply opacity penetrates into animality, points to the appropriate response to this objection.

According to Donald Davidson, "The intensionality we make so much of in the attribution of thoughts is very hard to make much of when speech is not present."[1] There is no point in denying or underplaying Davidson's correctness in this claim. There are some contents that acts cannot have unless the agent of the act is capable of speech. As Davidson argues, there is no point in trying to determine whether even an animal as intelligent as a dog goes to the door to greet its master or the president of the bank (who is its master) or Mr. Smith (who is that same master).[2] There is no point in raising this issue because in the absence of the power of speech, no possible evidence could count for or against assigning any of these goals to the exclusion of the others. Where there is no possibility of assigning a goal to an act, there is no possibility of assigning a content to the belief or desire that motivates the act. We can't intelligibly say whether the dog goes to the door because it believes that its master is home and wants to greet him or whether the dog goes to the door because it believes that Mr. Smith is home and wants to greet *him*. It makes no sense to assign *any* of these definite contents to the beliefs or desires of the dog.

Nonverbal animals are thus incapable of having certain kinds of thoughts with certain kinds of contents. Intensionality depends on distinctions, so if an animal is incapable of displaying the ability to make certain distinctions in its behavior, it is an error for us to assign the corresponding contents to its thoughts. As Davidson says, "The intensionality we make so much of in the attribution of thoughts is very hard *to make much of* when speech is not present." The contents of the intentional states of nonverbal animals are, at best, hopelessly vague when compared with the sorts of intentional states that verbal animals can and do have. Any theory of intentionality must show how it is possible to have thoughts with the kind of contents that our thoughts have.

But to say that it is difficult to make much of the intensionality of animal thoughts is not to say that we can make nothing of it. Indeed, as we will presently see, even the goal-directed behavior of nonrational animals displays *some* intensionality. When one overlooks or minimalizes this fact, one also overlooks the way in which the definite contents of our own thoughts are rooted in and made possible by the much vaguer contents of the intentional states of our animal cousins.

Consider the case of the poor frog that can be tricked into hunting and eating lead pellets that are shot across its field of vision, a bit of animal behavior made familiar by the philosophical literature concerning intentionality. As it turns out, frogs will shoot their tongues out at—and capture and ingest—lead pellets that have been made to float before their eyes, just as they will shoot their tongues out at, capture, and ingest flies that float before their eyes. Whatever the causal mechanisms that trigger this sort of tongue-shooting behavior in frogs are, they are triggered by the presence of dark floating objects of a certain size that move across the frog's field of vision in certain ways. In the usual environment in which frogs find themselves, only flies and other flying insects have this appearance and move in this way; such insects are prime food for frogs, and these triggering mechanisms serve frogs well. But just like our poor wasp Fran, who is tricked into a vicious circle of behavior in an experimental situation, frogs can be tricked into the act of eating lead, a behavior that I suspect is not very beneficial to frogs.

How should we interpret this case? The case rose to prominence because it points to a problem for models of intentionality that link the content of an internal state with the causal role of that state. Assume for the moment a simple form of causal-role functionalism. If an internal state of a frog, which when combined with the frog's hunger causes the hunting and tongue-flicking behavior, is caused by both the movement of flies and the movement of lead pellets, then, all other things being equal, why shouldn't that internal state be thought by the causal-role functionalist to have the content "that a fly or lead pellet is floating by"? But if the content of a belief state is always specified by a disjunction of all factors that might cause that sort of state, how could the belief ever be false? Since it is essential to beliefs that they can be false, this would seem to suggest that there is something wrong with the view that the contents of beliefs are fixed by their causal roles.

The pragmatic view has no particular problem with this case. Indeed, there is a sense in which that view is designed to dissolve this sort of problem. First, note that frogs' susceptibility to trickery gives us reason to believe that they are not rational and do not have reasons of their own for doing what they do, even though there are of course good reasons for them to act in certain ways. (Those reasons are "free floating," or "mother nature's reasons," to use some of Daniel Dennett's terminology.) Nevertheless, there is a clear sense in which the tongue-flicking behavior of the frog has a goal or set of goals. The frog flicks its tongue in order to catch an insect so as to make it a meal so that it can be nourished. The entire nested sequence of goals is anchored in the biological fact that the agent is an organism of a certain sort that has the intrinsic goal of nourishing itself by ingesting flying insects, and this anchor specifies that the frog hunts insects and not lead pellets. When the frog is tricked into eating lead, its act can be evaluated straightforwardly. Since the ultimate point of the act is to nourish the frog, and the frog is not nourished by lead, the act in which the frog catches the lead must be judged a failure.[3]

In the present context, what is interesting about this short frog story has little to do with the intentional contents of any supposed beliefs of the frog. What is arresting about the story is the fact that the frog's actual behavior can be normatively assessed for success or failure. The *goal* of the frog's act is to eat *food,* that is, organic matter that will nourish the frog. But all that results from the tongue flick is the ingestion of lead. So the tongue flick failed to realize its goal. This normative evaluability of goal-directed acts signals a further crucial fact concerning such acts, however. Because goal-directed acts can be normatively evaluated, the goals of such acts also exhibit opacity. How one *describes* the possible state of affairs that one asserts as the goal of a goal-directed act makes a difference to the truth of that assertion.

Whenever a frog captures a fly that is flying past, it is also capturing a dark object that is crossing its field of vision. It is doing so because any fly that flies past *is also* a dark object. It is for this reason that evolution designed the dark-object-detecting mechanisms that trigger the frog's tongue flicks: in usual frog environments, if you capture a dark flying thing, you also capture an edible insect. This is also why it is so easy to trick the frog into ingesting lead. Be all this as it may, however, when a frog acts so as to capture a fly, it is acting in order to catch an edible insect and not a dark object crossing its field of vision, even though the insect is such an object: the frog can catch a dark flying object (for example, a lead pellet) and fail to realize its goal, but it can't capture a fly and fail to reach its goal.

Although I have described the frog's situation so that its act has the goal of catching an edible insect and not a dark flying thing (even when the insect *is* a dark flying thing), the nested character of goal attributions allows an alternative description of the opacity involved here. One might say that the frog's act *does* have the goal of catching the dark flying thing, so as to get a meal, but when the dark flying thing is, as it happens, a lead pellet, the act doesn't have the goal of catching a lead pellet, even though the dark flying thing is a lead pellet. Needless to say, the same opacity is revealed in this description of the frog's act as in the other.[4]

Frogs may lie somewhere between wasps and birds when it comes to rationality, but there are good reasons to suspect that they are not especially rational. But when a frog flicks its tongue in the direction of a fly buzzing past its head in the course of a routine of hunting behavior, what the frog does has the goal of catching the fly going across its field of vision and not the goal of catching the flying dark object going across its field of vision, even though in this case the very same object uniquely satisfies both descriptions. That the frog's act has the first goal—and not the second—is indicated by the normative character of the specification of the goal. When an experimenter replaces the fly in the frog's field of vision with another, nonedible flying dark object (for example, a lead pellet), she tricks the frog into repeating the hunting behavior, right down to the tongue flick. In this case, the act does not achieve its goal, because the act makes teleological sense only as part of a pattern of behavior that serves to nourish the frog: ingesting lead pellets doesn't serve that ultimate end, whereas ingesting flies does. In the original case, as well, the edibility of the fly is crucial to specifying the goal of the act, and otherwise accurate descriptions of the fly that omit this salient feature therefore fail to specify the goal of the tongue flick.[5]

Given the criterion of opacity, then, contexts that attribute goals to the behavior of nonrational agents must, as such, count as intentional, for these contexts are opaque. The goal of this frog's tongue flick is to catch that fly, not to catch this dark flying object, even though that fly *is* this dark flying object. So the goal of this behavior has the content that the frog eat an edible insect but does not have the content that the frog ingest a dark object. This content—and this opacity—is entirely independent of whether the frog is rational and whether it has any beliefs and desires at all. Once one commits oneself to the existence of nonrational goal-directed behavior, one also commits oneself to the intentional content of that behavior.

Recognizing the degree to which nonrational goal attributions establish intensional contexts, and thus rely on intentional idioms, in no way undercuts the fact that the contents of the acts of nonrational animals are enormously vague. When a frog flicks out its tongue at a dark flying thing, which also happens to be a fly, the act does not have the goal of capturing a dark flying thing, even though what the act aims to catch *is* a dark flying thing. We know this because, even though this act succeeds when it captures *this* dark flying thing, it would have failed if this dark flying thing had not been edible. But there is simply no answer to the question of whether the aim of the act is to catch a fly or an edible flying insect or an edible insect or what have you.

Part of this vagueness is rooted in the source of the content of the act. The act has a content only in virtue of its having a goal. But the goal of the act is fixed by its being an instance of a stereotypical part of a stereotypical routine of behavior through which frogs of this type typically nourish themselves. That this act derives its goal from the part it plays in frog-nourishing behavior guarantees that it will be a failure if it results in the frog's ingesting a lead pellet and thus guarantees that any description, such as "dark flying thing," that fails to distinguish between edible insects and lead pellets will fail to adequately express the goal of the act. But because the goal of the act is just to nourish the frog, *any* description that distinguishes between results that involve the frog's being nourished and those that don't will be just as good. The *only* thing that distinguishes between dark flying things that nourish frogs, such as flies, and dark flying things that don't, such as lead pellets, is the *frog's needs*. The *frog* can't make that distinction. And since the frog's needs distinguish only down to the level of the difference between the edible and the inedible, any description that goes that deep will be as good as any other to characterize the goal of the act.

The vagueness of the contents of the goal-directed acts of nonrational agents is thus related to the fact that it is *only* those acts that have content. The frog doesn't intend to eat what it aims at, any more than the wasp intends to check for predators. For either to intend to do those things would require it to have beliefs and desires, but, insofar as the behavior is nonrational, neither has such intentional states.

The goal of an act is a holistic property of the act, and, in the case of nonrational behavior, the relevant whole is the stereotypical behavior that a type of animal agent uses in the course of achieving its fixed organic ends of survival and reproduction. Rational agents, on the other hand, are capable of novelty, flexibility, and versatility of behavior in response to changing environmental circumstances. So while such agents retain their fixed organic ends of survival and reproduction, the manner in which they achieve these goals, when they do achieve them, varies as a function of differences *within* each rational agent. The fact that behavior counts as rational by being mostly successful at attaining the agent's ultimate goals in changing environments implies that the states of the individual agent that explain this rational behavior must themselves satisfy a variety of holistic normative constraints that together help to fix the *content* of these states. In effect, the novelty and versatility of the behavior demand that the stable specieswide holistic pattern of behavior that provides the context in which nonrational acts have goals be replaced with an idiosyncratic, changing, holistic pattern of the individual rational agent's behavior, beliefs, and desires.

Individual nonverbal animals who act rationally act for reasons of their own. Such animals are capable of learning to make distinctions within their environment that nonrational animals cannot make. This, in turn, allows the possibility that these animals can have beliefs and desires whose contents are less vague than are the contents of the goal-directed acts of nonrational animals.

Consider, for example, a young cat. The description is taken directly from Alasdair MacIntyre. "[W]hen a young cat first encounters a shrew, it characteristically and generally treats the shrew exactly as it would a mouse. That is, it tries to catch it, and, if it does so, plays with it, kills it, partly skins it, and eats some. That cat will then become violently ill. Thereafter it will leave shrews severely alone. It now distinguishes between shrews and mice. What its actions show it to believe about shrews is no longer the same as what its actions show it to believe about mice."[6] Not all cats avoid shrews. We account for the actions of cats that have learned to avoid shrews by appealing to their beliefs and desires. But how ought we to characterize those beliefs and desires? It would be wrong to say that the cat has beliefs and desires whose contents concern *shrews*. After all, for a cat to intend something *as* a shrew, it would need to intend that thing as an animal of a certain sort, and it is anything but clear that we have any reason to think that cats can intend anything as an animal, let alone as an animal of a certain sort. Nevertheless, *the cat* does come to distinguish mice from shrews and to display this ability in its behavior, a trick that frogs can never accomplish vis à vis flies and lead pellets. *The cat* has a reason of its own for not eating animals that we class as shrews, in addition to the reason for cats in general not to eat animals that we class as shrews.

The contents of the internal states of the cat that motivate its behavior are *less* vague than are the contents of the frog's acts. The cat that has learned about shrews treats those things that we class as shrews differently than it treats all other small furry things. So the beliefs and desires that we attribute to the cat to explain her actions aren't about the small furry thing that just passed by *as* a small furry thing, whatever else they might be about. This is true even though that shrew *is* a small furry thing that has just passed by. Indeed, insofar as the cat who has learned about shrews comes to act differently toward shrews than it does toward any other class of things, there is some point in saying that the cat is not even intending that shrew that walked by as an inedible small furry thing, even though, of course, that shrew is (for a cat) an inedible small furry thing.[7]

Having said all of this, however, one must still admit that the contents of the beliefs and desires of even relatively rational animals, such as cats and dogs, remain quite vague in comparison with the contents of our human beliefs and desires. Even though cats are capable of distinguishing individuals, just as dogs are, there is no point to the question of whether the cat is bent on cozying up to its master or Mr. Smith or the banker or anyone else.

So animal instrumental rationality only partially closes the gap between the seemingly ineliminable vagueness of animal intentionality and the crisp definiteness of our intentional life. As it turns out, one of the ways to see how the two are related, and to see how the definite kind of intentionality can arise out of the indefinite, is to consider a second objection to the pragmatic view of intentional mental states.

## 3. REFLECTION AND PRACTICAL RATIONALITY

Philosophers who have attempted to understand human reason have fallen into two quite distinct traditions. One of these traditions has emphasized that an agent's rationality involves the ability to develop appropriate means to achieve specific ends in a variety of different circumstances. These ends, in turn, are usually thought to be definite states of affairs. So for this tradition, an agent is rational just to the extent that it is good at figuring out, in a variety of different situations, how to bring about some state of affairs that it aims for. This tradition divides on the question of how best to characterize the goals toward which rational animals aim. For some, those ends are prescribed by the nature of the animals involved. So in this view, it might be natural for porpoises to act with the goal of eating fish, and a porpoise would count as rational if it could invent novel ways to catch and eat fish on the basis of its experience of fish behavior. For others, the ends of rational action are fixed by the desire of the individual agent. Such philosophers hold that reason is, and ought to be, the servant of desire.

It seems obvious that the pragmatic view of instrumental rationality borrows elements from both wings of that tradition. In this view, rational animals aim to achieve ends that are fixed by the contents of their desires. Those desires themselves can be evaluated for rationality on a basis provided by the fixed organic nature of the organism. Sometimes, tits and dolphins act on false beliefs and inappropriate desires and attempt to achieve proximate goals that, in fact, are inimical to their ultimate ends. Furthermore, in the view presented here, all definite states of affairs aimed at by rational animals, both the ends fixed by nature and the proximate ends dependent on the desires of individuals, are rooted in the ultimate end of all organic life, that there be a continuation of the pattern of the activity of the life itself. Finally, however, in this pragmatic view, all instrumental rationality remains a matter of finding appropriate means to achieve definite ends.

The second philosophical tradition attempts to understand *human* rationality in a very different way. Although this alternative tradition is complex, it is convenient to focus on the part that includes Kant and his followers.

According to Kant, there are two crucial respects in which human rationality is distinguished from all animal analogues. First, Kant suggests that while everything in nature acts according to rule, only humanly rational creatures act according to a concept of rule, or principle: "[E]verything in nature works according to laws, [but] only a rational being has the capacity of acting according to the conception of law, i.e., according to principles."[8]

Second, Kant and his followers hold that human rationality is distinguished by the fact that human agents are reflective. The claim is best summed up by the contemporary Kantian Christine Korsgaard: "The human mind is self-conscious in the sense that it is essentially reflective."[9] One can get a sense of what it means for a mind to be reflective in Korsgaard's sense from the distinction she draws between human mental states and animal mental states. An animal's attention, she tells us, is "fixed on the world. Its perceptions are its beliefs and its desires are its will. It is engaged in conscious activities, but it is not conscious *of* them." We humans, on the other hand, "turn our attention on our perceptions and desires themselves, on our own mental activities, and we are conscious *of* them. That is why we can think *about* them."[10]

How are we to understand these twin claims regarding human rationality, and how are they related? The key to understanding this way of approaching human reason is to appreciate the role assigned by these thinkers to the capacity for reflection. As Korsgaard says, some agents—surely including adult humans—are capable not only of thinking about the world but also of thinking about their own, and others', thoughts about the world. We are capable of second- (and third- and fourth-) order intentions. We not only believe that there is a leopard in the bush, but we also believe that the leopard wants to eat us; we not only want ice cream, but we also believe that we want ice cream. I not only want to eat that banana, but I also believe that you believe that I want to eat that banana, and I want you to believe that I don't believe that that banana is over there.

The recognition that humans at least, are capable of higher-order intentions is important from a number of different angles. In the Kantian tradition, the fact of second-order intentional states directed toward our own beliefs and desires opens up a gap between the beliefs and desires that characterize our own unreflected selves and our considered, higher-order, intentions. Each of us can intend the fact that we, for example, desire ice cream, and this intention is distinct from the desire itself. The fact that we are at least instrumentally rational creatures ensures the possibility that we may have reason to come to have a second-order desire that conflicts with our first-order desire for ice cream, and, since we are rational, this ensures that often we will come to have this higher-order desire. So, for example, we might come to desire that we *not* desire ice cream, while still desiring ice cream. But in that case, it would seem that the two desires enter into a conflict that cannot be resolved solely on the basis of either desire. I have both an instrumental reason to act so as to get ice cream and an instrumental reason to act so as to no longer have that instrumental reason, which might very well give me reason to act so as to not get the ice cream.[11]

This potential conflict of desires in creatures endowed with second-order intentional states introduces a problem and a possibility. The problem is simple: What ought we to do, and how are we to decide what we ought to do? If our desires conflict, then we can't simply appeal to what we want to determine what we should act to achieve. How can we determine this? This is what Korsgaard calls "the problem of normativity":

> [Our essentially reflective nature] sets us a problem no other animal has. It is the problem of the normative. For our capacity to

> turn our attention on to our own mental activities is also a capacity to distance ourselves from them, and to call them into question. I perceive, and I find myself with a powerful impulse to believe. But I back up and bring that impulse into view and then I have a certain distance. Now the impulse doesn't dominate me and now I have a problem. Shall I believe? Is this perception really a *reason* to believe? I desire and I find myself with a powerful impulse to act. But I back up and bring that impulse into view and then I have a certain distance. Now the impulse doesn't dominate me and now I have a problem. Shall I act? Is this desire really a *reason* to act? The reflective mind cannot settle for perception and desire, not just as such. It needs a reason.[12]

Korsgaard's use of the word "reason" in the quote above indicates the possibility that this tradition finds in the problem. Korsgaard, like Kant before her, holds that the agent in this situation requires a justification for action that is in no way rooted in the particular, already determinate, ends of that agent. Since the agent finds itself at war with itself, the fact that it is an agent of some determinate sort is no longer relevant to the determination of what the agent will do. Inorganic beings simply will do what the descriptive laws that govern beings of that sort prescribe. A lead ball will fall in the earth's atmosphere, and a helium-filled balloon will rise. They act according to the laws that cover beings of their type. Organisms will do what organisms of that type always do in a given situation. Beyond that, even instrumentally rational animals can be counted on to act in their perceived self-interests when those perceptions generally correspond with the actual interests of the organism. All of nature, as Kant says, works according to law.

The Kantian tradition holds that the normative problem for reflective minds arises out of the fact that such individuals recognize that they must be agents but also recognize that they naturally belong to no *type*. Since reflective agents are potentially alienated from every impulse that could type them, they stand under no law or rule that governs agents of any natural type. In fact, however, there is a type to which reflective agents belong. They are reflective agents, that is, agents who must be agents regardless of the fact that they belong to no natural type. To act, an entity must be an agent; to be an agent is to act according to a law that governs actions of agents of a type; and the only type to which reflective agents belong is the type "reflective agent," that is, an agent that is not determined by any natural fact about itself. As Kant famously argued, it follows from these considerations that the law under which a reflective agent must act in order to count as an agent is a law that the agent has given to itself (since it can't be a law that the agent falls under in virtue of some determinate fact about that agent). All that is required of this self-given law is that it be a law (since any further content would need to be borrowed from some natural determination of the agent, a move that is disallowed).

There is a crucial difference between a law under which an entity stands in virtue of some natural fact about that entity and a law under which an entity stands in virtue of having imposed a law on itself. The first type of entity stands under the law regardless of whether it recognizes that it stands under that law. If the entity gives the law to itself, it stands under the law only insofar as it acknowledges that it stands under the law. In that case, the second-order intention in which the reflective being *accepts* the belief that it is a reflective being that acts according to a self-imposed law plays a critical role in determining what that being is and what that being does. That is, the entity stands under the law only insofar as it *commits* itself to the law, to use Korsgaard's term. This commitment serves as a *premise* from which the agent can and does infer what in particular it is to do. Since I accept that I am a self-determining agent who must, to be so, act according to law, I infer that I should act in such a way that what determines my act could be a self-imposed law for any reflective agent.

The pattern for the explanation of the acts of a reflective agent is thus quite different from the pattern for the explanation of the acts of other, that is, natural, agents. Natural agents are caused to act as they do by their natures. Reflective agents act as they do because they accept reasons from which they can infer actions that are appropriate given those reasons. These reasons have the form of laws. That is, they are universal generalizations concerning what always is, or ought to be, the case. To accept such a reason is for an agent to come to *believe that* if some type of initial condition is met, some other type of condition will, or ought to be, met. So when the agent comes to believe that some instance of the antecedent has occurred, the agent *infers* the consequent. In the case of theoretical reason, the agent infers from its belief in the general law that it ought to come to believe that the consequent is the case. In the case of practical reason, the agent infers that it should act so as to bring about the consequent. But in both cases, the explanation for what the agent does passes through the agent's *beliefs* regarding general laws or principles and the agent's syllogistic drawing of a conclusion from those principles. To reiterate Kant's words, "[O]nly a rational being has the capacity of acting according to the conception of law, i.e., according to principles." The twentieth-century Kantian Wilfrid Sellars put a slightly different slant on this same point regarding human rationality when he said that we humans "follow rules" and then went on to say that "if the expression 'following a rule' is taken seriously, and is not weakened beyond all recognition into the bare notion of uniformity . . . then it is knowledge or belief that the circumstances are of a certain kind, and not the mere fact that they *are* of this kind, which contributes to bringing about the action."[13]

For a Kantian, the possibility of higher-order intentions both demands and allows the possibility of "acting according to principles" or "following a rule." When an agent follows a rule, what it does is explained by appeal to a process in the agent in which that agent accepts some rule, comes to believe that the antecedent of that rule is satisfied, and infers an action from these first two steps. This process can be conscious or unconscious, implicit or explicit. But what is crucial for human rationality is the ability to be motivated to act by the combination of a belief in a general proposition and the act of inferring a conclusion from it. If an instrumentally rational agent has reasons of its own for what it does, the agent that is *practically* rational in this Kantian sense can be said to act in a certain way because of its *recognition* that it has reason to do so.

This style of rationality, if it exists, contrasts markedly with any example of merely instrumental rationality, no matter how sophisticated. The merely instrumentally rational agent is always motivated by its beliefs about what is the case in the world and its desire to bring about some state of affairs. As opposed to merely goal-directed organisms, the instrumentally rational agent has reasons of its own, but those reasons always involve the agent's desiring to bring about some determinate state of affairs. There is a sense, however, well brought out by Kant, in which the practically rational agent is not motivated by desire at all. What is crucial to how the agent acts is which *principles* the agent *accepts,* which general propositions the agent recognizes as authoritative.

In the central case, that of determining what the agent itself is to do (theoretical reason is just a special case in which what is determined is what the agent is to come to believe), the general principle that the agent accepts and from which it infers its action concerns what it is that *every* agent, or every agent of the agent's type, is to do. So, for example, when a practically rational agent who also happens to be a mother acts to comfort her child, she does this because she accepts some principle of the form "Mothers comfort their children" and infers from the fact that she is the mother of this child that she will comfort it. If the mother is entirely practically rational, then she will do this whether she wants to or not because, as practically rational, her motivations are not determined by her desires but by her principles and her reasoning. Indeed, there are instances in which it makes perfectly good evolutionary sense for even an instrumentally rational mother not to want to take care of her young—for example, when doing so would, because of maternal exhaustion or malnutrition, endanger the mother or her other children. But the practically rational mother would, if she accepts the principle of comforting, act to comfort the child even when she doesn't want to. Contrast this explanation with an instrumentally rational explanation of the same sort of act. In that case, what the mother does—comforts her young—occurs because she wants her child to prosper and believes that comforting the child will help the child to prosper. That, in general, the mother wants this outcome is guaranteed by the fact that she is an organism of a certain sort that has the goal of reproducing and the fact that, as rational, she generally wants what she should, given her ultimate fixed organic ends. (The explanation for why there are mostly mothers of this type, of course, is evolutionary.) But none of these facts about the instrumentally rational mother guarantees that she will *always* want to comfort her young. In those cases, she simply will not act with this goal because she is motivated *only* by her desire.

Now when a practically rational agent is motivated to act by its "following of a rule," it would be correct to say that the agent is acting *for the sake of being a certain sort of agent* rather than acting *in order to* bring about some end, to borrow a distinction from Heidegger. The mother who does not want to comfort her child but does so anyway because she acts on principle, acts as she does so that she will be an instance of a certain type of agent—a mother—for, after all, she accepts the general principle that mothers comfort their young.

Human beings are capable of this sort of practical reason. Without this capacity, it is hard to understand how the kinds of rational action exemplified by the ascetic or the martyr could be possible. It is only the ability to act on principle that allows us to see how any agent could have a *reason* to act in these ways. But if it is possible for human beings to act on such principles, then doesn't this call into question the account of rationality that I have offered in the previous chapters? For that account asserts that it is a necessary condition on any agent's acting for a reason that the agent be an organism of a certain sort and asserts that all reasons are grounded in the fixed, organic ends of organisms. Practical reasoning of the kind we have been discussing seems to depend on the alienation of human beings from their merely animal natures.

## 4. THE EVOLUTION OF PRACTICAL RATIONALITY

How are we to understand, from the biological perspective of this book, the human capacity for practical reasoning in the Kantian style? And how is it possible for human beings to have thoughts with such incredibly (from the biological perspective) precise contents? The answer to the first question is simply that, given the kind of *organism* we are, we have good *instrumental* reasons to come to adopt the *practical* form of reasoning in addition to the instrumental. And, since we are *very* instrumentally rational, and we generally do what we have good instrumental reasons for doing, we have come to adopt the practical form of reasoning. This adoption, in turn, allows the possibility for developing articulated language, which is the condition sine qua non for having thoughts with highly determinate contents.

To help the reader grasp the character of this solution to these problems, I have found it convenient to briefly discuss some facts about our own evolutionary situation and inheritance. I do this even though I'm afraid that, to some philosophers, this discussion will seem a merely irrelevant speculative fantasy. The point of this discussion, however, does not depend on the details of this evolutionary story. That point is that human beings are organisms that make their living by dwelling together in a highly flexible, yet highly stable, socially differentiated group. This sort of group is distinguished by the fact that the success of the group, and thus of each of the members of the group, depends on the ability of individuals to perform certain socially prescribed tasks in certain ways, even though those individuals are not equipped by nature or evolution to perform those tasks. This divorce between what we, as individuals, want to do because of our biological inheritance and what we need to do as members of our social group, allows the possibility of a conflict of desires. Given this social fact of our biology, there is a good reason for human beings to develop the ability to act on principle in the Kantian sense.

We human beings are organisms. As with every other kind of organism, we need to make a living. And as with every other kind of organism, the kind of living we can make is limited by the resources and constraints supplied by our morphology and inheritance.

Although there is quite a bit of scientific conflict regarding the details of that inheritance, certain central aspects of our situation are clear.[14] Whether our immediate ancestors were essentially groups of related and bonded females with male hangers-on who made a living by gathering vegetable resources (like many monkey species) or essentially groups of bonded and cooperating males who hunted together, protected a common territory, and shared food with and promiscuously mated with a group of somewhat solitary, itinerant females (like contemporary chimpanzees), the details of our anatomy and our developmental arc make it certain that these ancestors lived cooperatively and socially, as we do today. Human infants are born at an exceedingly early stage of their physical and mental development, and they have an exceedingly long childhood period. It is virtually inconceivable that single mothers, operating outside of a social context, could successfully raise such young. In addition, our closest kin, the common chimps and bonobos, are exceedingly social animals who share food and live in "fission-fusion groups" (groups that stay together for extended periods of time as an identifiable community, but that (1) separate into smaller subgroups or individuals during the day for the sake of food acquisition and (2) tend to divide and reform with elements of other such groups). Solitary primates of our sort have a very difficult time of it indeed if they find themselves exiled from their communities. For chimps, bonobos, and humans, it is a condition of making a living at all that we live together in social groups. Fish got to swim, birds got to fly, and we got to be social.

Many kinds of animals live together in groups, of course. These groups come in many different varieties, from the beehives and ant heaps of the social insects, through the herds of mammal herbivores, to the packs and troops of dogs and monkeys. There are several distinctive, if not unique, features of the kinds of communities that are a condition on chimpanzee, and human, life, however.

Chimps and humans share with dogs and monkeys a tendency to form structured "dominance-hierarchical" societies in which each individual has a place in a structured order of submission and dominance. There are chimp—and high school—pack leaders just as surely as there are canine pack leaders. There are good reasons for groups of social animals to be organized in this way, although they are surely neither reasons left by evolution to be idiosyncratic reasons of an instrumentally rational individual's own nor reasons that practically rational agent chimps treat as principles. Instead, they have to do with the need for quick, focused, cooperative action in which each member of the group plays a particular, but distinctive, role in an articulated group effort. As long as the style of life of a social organism demands only that all of the members of the group act in a similar fashion (as is the case in herd animals), the group can get along with one leader and many undifferentiated followers. But when each member of the group must act in a different, but coordinated, fashion for the group to survive, it is helpful if there is some simple, yet flexible, mechanism for assigning such parts. Hereditary assignment of roles in a system of hierarchy-dominance, as is evident in many monkey species, is an obvious solution to this problem.

Humans and chimps, however, go even further in this same direction for the organization of their social groups than do dogs and monkeys. Both the occupants of various social positions and the details of the behavior that is demanded of the occupants of those positions tend to change relatively rapidly and thus are fixed less by inheritance than by interaction among the specific members of a group at given a time. In many monkey species, one can predict quite accurately from the sex, maternity, and birth order of an infant what position that infant will come to occupy in its group and how that infant will come to act toward, and with, other members of its group. That is, in these monkey societies, social roles are part of the inherited pattern of life that serves to fix the identity of the species and that can be used to teleologically predict and explain the behavior of individual members of the species. This is not the case in chimps. Both the position within the hierarchy that an individual comes to occupy and how that individual performs the role coordinated with that position are functions of the idiosyncratic personalities of the members of the group and the social negotiations that occur among those members. Among the common chimps, for example, most head males attain their position neither through size and strength nor through inheritance but by forming and sustaining alliances with other males in the group and by gaining the willing acquiescence of the females. These head males display many different styles of leadership, from the autocratic and tyrannical to the altruistic and cooperative. (Sound familiar?)

Once again, Mother Nature has good reasons why chimp and human societies should come to have this shape. From five to six million years ago, when chimps and humans were differentiating and rapidly evolving in Africa, the world was going through a drying and cooling period in which jungle was being replaced with savannah. To some extent, the chimps came to make their livings in the border zone between these two environments while humans quickly moved from the forest to the grass plains, an environment they were not yet evolved to cope with. As social animals, the only hope that we and the chimps had to survive in such a novel, rapidly changing, and variable environment was to develop a form of social organization that maximized the potential for innovative and flexible responses to changes in that environment, while maintaining the group's ability to work together in a coordinated fashion. In such societies, an extremely high premium is placed on the social intelligence of individuals, the ability to figure out what one's companions will want to do next and, given those desires, to design a course of behavior that will induce those companions to act in socially desirable ways. Mother Nature could hardly have designed a better way to determine who should come to lead these new kinds of societies than by making leadership dependent on the capacity to succeed in intragroup social negotiation, as it is in human and the chimp societies.

There are, however, several obvious and important differences between chimpanzee societies and human societies. Human societies are much larger and more stable than chimp societies.[15] Chimp communities tend to fission when they reach 120 or so members, sometimes leading to civil war and the killing of some of the former members of the troop. Chimps from different groups who have never personally met are highly unlikely to cooperate with one another on common projects. Human communities, on the other hand, can reach truly enormous size. These large human communities are in turn composed of a series of nested subgroups whose members might be totally unknown to each other. Nevertheless, these seemingly mutually isolated individuals in their different subcommunities find it remarkably easy to interact and cooperate with one another when occasion demands it. When I fly from London to Boston on a Canadian plane, I am unlikely to meet anyone I have ever encountered before. Nevertheless I can get on the plane with perfect confidence that a large number of people who are utter strangers to me will cooperate and perform highly specialized functions that, when taken together, will result in my getting to where I want to go. From the perspective of comparison with chimpanzee societies, this is an extremely surprising fact.

The difference in size and cooperative ability between human and chimp communities is associated with another stark difference. The roles allotted to the various members of a chimpanzee troop are themselves highly unstable when compared with those occupied by their human counterparts. In chimp societies, one's position in the hierarchy and one's interactions with the other members of the group vary as a function of one's alliances, and these in turn vary on a regular basis. Palace revolutions at the top and minor coups lower down are quite common in chimp society. So the roles occupied by the various members of a group tend to be fluid to a degree that is extraordinary when compared to the stability of the complex roles of pilot, flight attendant, engineer, and air traffic controller, each of which has its own duties, rights, and privileges, (as I was once told on attaining a particular status within a human hierarchy). This relative lack of stability in the occupation of chimpanzee roles makes it highly unlikely that chimp social roles would ever develop and ramify to the extent that human roles have, even were chimps to develop the necessary degree of instrumental rationality.

There is an excellent explanation for this inability of chimps to establish large and stable communities in which individuals occupy long-term, separate roles. The alliances through which individual chimps attain a given status are established and maintained through acts of reciprocal altruism among the members. For example, a higher-ranking male who wants the support of another male in a coming dominance contest with a head male might share some especially precious bit of food with his subordinate. A subordinate is highly unlikely to support a superior unless he can expect something in return. Altruism is essential to the formation and maintenance of alliance and thus essential to attaining and maintaining a given social position. But such altruism is apparently not easy for chimps; they have a tendency to hide, cheat, steal, and lie when it serves their immediate self-interest.[16] And when a chimp acts in that way, alliances dissolve and positions are lost.

That this pattern of behavior should be prevalent in a highly social, highly instrumentally rational animal like a chimp (or a human) is perfectly predictable. For an instrumentally rational animal, the whole *point* is for that animal itself to survive, prosper, and reproduce, and such an animal is likely to have beliefs and desires that help it to secure these ends. But chimps operate under a variety of social constraints on their ability to attain these ends, and these constraints pull in opposing directions. First, chimps must live together with their fellows and sisters in their community, and to some extent the community as a whole must prosper in order for the individual chimp to attain the goods that it needs. But within the group, those individuals who obtain a greater share of the goods of the community are in a better position than those who receive a lesser share. The share of goods—for example foods and matings—that an individual obtains will be a function of two opposing factors: one's position within the hierarchy and how well one *circumvents* the rules operating within the hierarchy. While in general those individuals who attain and maintain a higher status are more likely to survive and reproduce than those who occupy a lower position, if a lower-ranking male can "go on safari" (as the primatologists call the behavior of an amorous pair who go off into the bush by themselves, sometimes for an entire estrous cycle in the female) with a higher-ranking female (who herself might benefit from having the offspring of a younger but lower-ranking male) or if a female who finds an especially productive banana tree can keep it to herself, these individuals are more likely to survive and reproduce than are those who don't cheat. On the other hand, if one gets caught in cheating in these ways, it is likely to damage one's relationships and one's ability to attain one's ends. So from an instrumentally rational short-term perspective, it is *entirely rational* for chimps to act as if they are following the rule "Appear to act justly, but act unjustly if there is a good chance of getting away with it."

Individual selection and the most obvious dictates of instrumental reason (given the fact that it is instrumentally rational to turn on and to distrust an individual who has tricked you) thus favor a kind of behavior in chimps that inevitably leads to small, unstable, fluid societies. But given the territorial and aggressively hostile behavior of chimps toward *other* chimp communities, any chimp community that could figure out a way of growing larger and maintaining more stable and fixed alliances and social roles would have an enormous competitive and selective advantage against other chimp societies. That is, when it comes to competition between *groups* of chimps, the group that has a high proportion of members who do what they are supposed to do even when it is not in their immediate interests wins hands down, even though within each group the successful cheat does best.[17] Since it is far more important to any chimp that its group should do well than that it as an individual should do relatively better than the other members of its troop, it turns out that, in the long run, it is instrumentally rational for chimps—and for us—to have beliefs and desires that lead to following the rules of our community rather than cheating.

So the more instrumentally rational a chimp or a chimplike animal becomes, the more likely it is to develop some way of getting itself, in general, to follow the rules that are appropriate for its position in society rather than cheating for its own immediate benefit. One good way to obtain this result, of course, is to have an animal that is so constituted that it acts for the sake of instantiating some principle. That is, any animal that acts as its does because it has a reason to do so, in the sense that it believes that *anyone* of a certain type *should* act in this way and recognizes herself to be of that type, will be doing what is, in the long term, instrumentally rational by doing what is practically rational.

## 5. PRACTICAL REASON AND SOCIAL ROLES

The Kantian type of practical reasoning demands that an act be justified in a way different from the way an act is justified instrumentally. For an act to be justified instrumentally, it must be an act that, in the circumstances in which it occurs, brings about the goal for which the act is performed. Since an agent has reasons of its own that justify an act only if the agent is instrumentally rational (and an agent is instrumentally rational only if it acts out of its own beliefs and desires and mostly believes and desires what it ought, given its biological nature), for an act to be instrumentally justified generally involves that act's being one that brings about the satisfaction of the agent's desire. That is, it must bring about a certain set of consequences.

But the desire of the agent is irrelevant to the Kantian type of justification of an act. For this type of justification, the act must be one that arises out of an agent's belief that some principle holds and an inference that in order to instantiate this principle the agent must act in some definite way. Because the belief that the act is an instance of a certain *kind* of act provides the reason for action, the *consequences* of the act are irrelevant to the justification. Instrumentally, an act of maternal altruism is justified if it satisfies a maternal desire to help her child. Practically, an act of maternal altruism is justified if it is an instance of a practical rule of the form "Mothers ought to act so as to help their children," even if the mother has no desire to help and even if doing so brings about results that do no good in evolutionary terms. *That* human beings act in a practically rational way is justified instrumentally, but any given act that is practically rational need not be justified instrumentally in order to be an act that is rationally justified.

In the previous section, I argued that there are good evolutionary reasons for our developing the capacity to act on principle. As instrumentally rational animals, we mostly want what we should, and what *we* should want is to become an animal motivated to act on principles. Possessed of instrumental rationality, we have developed the second-order intention of wanting to become a practically rational creature. And since, as instrumentally rational, most of our projects are successful, it is mostly true that we *do* act on principle.

Because this is the case, it will frequently be possible to predict and explain mature human action through a process rather different from the one by which one predicts and explains the behavior of merely instrumentally rational agents. If one can identify the general principles that motivate the practically rational agent, one can assume that the agent has a distinctive type of desire. The instrumentally rational agent desires that some definite result be brought about. But for the practically rational agent, the consequences of an act are irrelevant to the motivation of the act. What the practically rational agent *wants* is that its act instantiate the rule that justifies it. This desire is *formal* in the sense that it is the same for *any* practically rational action. What the practically rational agent wants is to follow the rule, but the *content* of the rule, while relevant to whether it ought to be a rule (we will return to this below), is irrelevant to the agent's wanting to follow it. As Kant says, the practically rational agent is motivated by "respect for law" itself.[18]

Because this is the case, an observer needs to know that the agent *is* practically rational and to know which practical principles the agent accepts or is committed to in order to predict the actions of that agent. The *desire* of the agent drops out as a motivating factor, as the motivating desire to act on principle is already implicit in the fact that the agent is practically rational.

The practically rational agent is thus motivated to act by a respect for law as such. But, as with any other rational act, desire—even the desire to act on principle—does not motivate by itself. Respect for law motivates action only when it is combined with belief, the belief that some general principle is true. A brief reflection on the evolutionary background of practical reasoning reveals the unique character of the contents of these principles. What was valuable for our ancestors was to develop the habit of reliably performing the actions that went along with their specific positions and roles within their group. For instrumentally rational protohuman agents, these roles and these performances did not come naturally; the roles had to be learned through demonstration, attempt, and correction. Through this process, the individual learned both what was demanded from any occupant of a given role and that the individual was to act in a way that satisfied that role. That is, the individual came to believe that all mothers, for example, act in a certain way in a certain situation and that she, the individual herself, was to count as a mother. So the form of the practical inference is a syllogism in which the major premise has the structure "(I believe) all M's act in manner A in situation S," the minor premise has the structure "(I believe) I am an M in situation S," and the conclusion has the structure "(I believe) I do A." The agent infers her act from her belief regarding what constitutes a given role and her belief that she occupies that role.

In general, then, one should be able to infer the content of the principles that motivate practical activity from the distinctive roles to which a practically rational agent commits itself. Each such role is characterized by a distinctively different set of norms for appropriate behavior, and in order to come to occupy some such socially recognized position, an individual must learn how agents of that sort are to act in a whole variety of situations. Mothers, when acting as mothers, act as mothers act; carpenters, when performing the role of carpenters, act as carpenters act; and flight attendants act as flight attendants act, at least while they are doing their jobs. Agents who cease to be motivated by a correct understanding of a social role or who cease to identify themselves as an occupant of such a role, cease to count as agents of that sort. And this fact makes it possible for another individual or the agent itself to be able to make quite accurate predictions about human behavior based solely on an understanding of a social role and the apprehension that an agent occupies that role.

Since, however, the agent also recognizes itself as occupying other roles and, as an instrumentally rational agent, finds itself motivated by its idiosyncratic desires, the agent confronts a number of different options and so experiences the act as something it *ought* to do. Because the agent infers its act from its being a certain kind of agent, the agent experiences the imperative to act in the way demanded of that kind of agent as the imperative to *be* an agent of that kind. The mother ought to comfort her child because that is what mothers do and because she is committed to being a mother.[19]

The beliefs of the practically rational agent that it is an agent of some definite sort and that agents of that sort act in certain definite ways are holistic properties of the agent. The holistic context in which any agent can have such beliefs necessitates that the agent's overall behavior and the beliefs and desires that motivate that behavior must be related in ways that satisfy the normative constraints on rationality. In particular, the beliefs of the practically rational agent satisfy two such normative constraints. First, no agent has any beliefs at all unless *most* of those beliefs fit together in a way that is *instrumentally* rational in relation to the agent's organic ends. This is a requirement on having beliefs at all. Second, no agent can act in a practically rational way unless it mostly infers what it ought from its beliefs. The first constraint points the way toward an understanding of the way in which practical rationality is rooted in instrumental rationality. The second constraint points toward an understanding of the way in which it is possible for a practically rational agent to act, rationally, against its own biological interests.[20]

It is a necessary condition on practical rationality that the agent who acts on principle also be an organism with intrinsic ends, because no agent can have any beliefs at all unless *most* of its beliefs are instrumentally rational in relation to the agent's organic ends. The practically rational agent acts on its belief that some general proposition is true. The mother in the previous example believes that all mothers act to comfort their children when those children are in distress. But what it is for the mother to have this belief implies that she will act in certain ways, given her other beliefs and desires, and what it is for her to have this entire system of beliefs and desires is intelligible only in terms of her *mostly* acting in ways that are instrumentally rational to achieve her fixed biological ends. If the mother didn't engage in versatile novel behavior that is mostly successful at getting food for herself and her child, at avoiding predators, at navigating the difficult waters of social life in her group, there would simply be no point in claiming that she has any desires or beliefs at all. So just as instrumental rationality is a system of behavior that depends on and survives in a pool of teleological necessity and success, practical rationality is a system of behavior that depends on and survives in a pool of mostly instrumentally rational action.

But it doesn't follow from the fact that no agent can act on principle in a practically rational fashion unless most of what it does is instrumentally rational that everything the agent does on principle is itself instrumentally rational. To illustrate this point, it is helpful to return to the evolutionary situation.

Evolution has not equipped human beings with the ability to act in a practically rational manner. No infant acts out of principle. Evolution has, however, equipped us with the capacity to develop this ability. Not only are we born into a community, but we are also born with the seemingly innate, formal desire to mimic the behavior of those around us and to mold our behavior in accordance with the wishes of our companions. That this should be so makes a great deal of evolutionary sense. Our societies require of us the ability to occupy a position in a group that is organized in an articulated manner, so that the various members of the group can act in a differentiated but coordinated fashion in response to environmental pressures. But since those pressures are highly variable, just how the community will need to be organized and integrated, and just which socially prescribed roles will need to be filled, will itself vary. So it makes perfect sense for Mother Nature to design a creature with a natural predisposition to learn to cope in whatever community it finds itself by developing principled habits of thought and behavior through a process of communal education. That is, it makes sense that there should be an animal like us, that seems to have a predisposition to learn how to follow the rules but for whom the rules to be followed will vary, depending on the human environment it is born into.

If humans weren't this sort of animal, the phenomenon of human culture would be incomprehensible. Each generation inherits from the last a set of roles and expectations organized according to an implicit set of practices, which can, to some degree, be reflectively formulated as principles. But these practices are applied in continually changing circumstances. As a consequence, there is both a high degree of stability and a remarkable amount of adaptability built into the traditions passed down through the generations. Isolation of communities inevitably gives rise to local variations in the social practices that (literally) inform the manner of life of different historical societies and give substance and content to the roles that are available to and prescribed for the members of those communities. We have a natural tendency to learn how to live by following the rules, but the rules we come to live by vary with time and place.

To become a mature human, one needs to do more than physically develop. One must learn to act as one ought to act in one's own community and in one's own role. One must learn to act as *one* acts, as Heidegger puts this point.[21] So every mature human being comes equipped with a set of principles and practices that she accepts and follows, a set of practices that together constitute her own self-understanding. But for a practically rational agent, these practices and principles that have come to be, as it were, a second nature, can never be the last word. As rational, such agents mostly believe what they ought to believe, given their other beliefs. This implies that mature human beings will continually change so as to consistently develop the consequences of their current beliefs and to overcome any inconsistencies among those beliefs. That is, as practically rational, human agents are always capable of subjecting their inherited beliefs and social roles to critique and development. As the tradition stemming from Kant has always recognized, it is possible for a practically rational agent to raise the question whether any principle that she finds herself holding is *justified*. This, in turn, allows the possibility that some human beings will come to hold and act on principles that are quite distant indeed from any principle that was justifiable from a purely biological perspective.

The unique biological requirements of our human ancestors set up a situation in which there was an instrumental justification for the development of the ability to act in a practically rational way, on principle. At least initially, the principles that our ancestors acted on must have made good biological sense in the context in which they operated. It is a good bet that in the long run, most of the principles that govern our lives, taken together, will also make good sense, given our organic and social requirements. Because human agents are practically rational, we will always drive toward the consistency and inferential closure of any system of principles that we find to motivate our behavior. This allows the possibility that we can come to be motivated to act on principles that neither can be derived from nor are consistent with the principles that would be justified on purely biologically instrumental grounds. For, while the entire *system* of action on principle has its justification in the requirements of our biological nature, no single principle within that system is justified by appeal to those requirements. Such principles are justified only—if they are justified at all—by appeal to other principles within that same system of practical reason.

The principle of charity guarantees only that most of what an agent does will satisfy whatever the charitable constraint is, not that an agent will satisfy that constraint in any given case. Even while the suicidal bomber is acting, rationally, with the end of killing himself, the very fact that he is acting rationally with that end guarantees that he can be *very* good at staying alive in varying novel circumstances up until the point at which he succeeds in killing himself. The fact that all of this instrumental rationality is in service of an end unjustified by biology in no way negates the fact that the agent could not have that end at all unless he were an instrumentally rational biological agent.

We can thus see how it is possible for human beings to act rationally in ways that are inimical to their intrinsic goals as organisms, even though it is a necessary condition on any agent's acting for a reason that the agent be an organism of a certain sort and that all reasons are grounded in the fixed, organic ends of organisms. When a martyr dies for a cause or a religious devotee takes a vow of chastity, what that person does can be rational in the sense that it is an act inferable from some principle that she accepts and that is derivable from some set of other principles that she already holds and that define her practical identity. But it is a necessary condition on there being this kind of practical rationality that the agent that acts on principle also be an organism with intrinsic ends, because no agent can have any beliefs at all unless *most* of its beliefs are instrumentally rational given the agent's intrinsic ends. Since there are no reasons without instrumental reasons, in this sense, all reasons are grounded in the intrinsic goals of organisms. But it doesn't follow from this that all rational acts are rational because they are instrumental to the achievement of organic ends.

## 6. TOOLS, LANGUAGE, AND DEFINITE CONTENT

The human capacity to act practically operates in an instrumental context, as we have seen, and this guarantees that many human roles will themselves have an instrumental point. It is a good thing that human beings come to be practically motivated because human beings make a living by operating together in a socially differentiated fashion, and acting on principle is a good way to assure that different individuals will reliably perform their roles. But for that reason, those roles will at least initially embody types of activity that have general instrumental utility. All animals must achieve certain ends in order to stay alive. Instrumentally rational animals achieve those ends by responding in a versatile fashion in shifting environments. Human animals attain those ends by parceling out different necessary ends as the tasks to be accomplished by different members of the group and by developing, over generations, distinctive styles of action that are well-suited to accomplish those ends. To educate an individual to act on principle is, typically, to teach that individual to be motivated to act so as to bring about a given end by acting in one of those distinctive styles. To be a carpenter is to act as carpenters act for the reasons that carpenters act, but carpenters act as they do because how carpenters act is instrumental to building socially useful things out of wood. An individual acts as she does so that she can be a carpenter, but to be a carpenter is, in part, to be someone who acts in a socially predetermined fashion with the goal of making things out of wood.[22]

Following Martin Heidegger, we can thus distinguish these two intertwined levels of human teleology by saying that, as human, we act both in order to realize determinate ends and for the sake of realizing certain possibilities of human existence. The doctor acts in socially prescribed ways and uses socially defined tools in order to cure her patient and at the same time acts in socially prescribed ways and uses socially defined tools for the sake of her being a doctor. To be a doctor is to engage in the practices that define the role of doctor for the reason that these are the right things for doctors to do. For that very reason, to do this is also to act in order to bring about a cure.

The practical activity of human beings is also typically goal-directed, and all goal-directed activity is, as we have repeatedly seen, holistically related activity that is mostly successful at achieving its end. So the various roles that are open to human agency are themselves defined by sets of interrelated procedures and practices for attaining the ends characteristic of a role, which an apprentice must master in order to occupy the role. Whether these procedures and practices can be formulated into explicit rules by the agents—or even whether these practices are formalizable in this fashion at all—is mostly irrelevant to the actual role. What is important is that the novice come to be motivated to act in the ways demanded by the role *because* they are demanded by the role and that the system of behavior that is specified by the role in fact be instrumental to the achievement of the ends of the role. As long as a carpenter can be counted on to build things well by reliably following correct carpentry procedures, it doesn't matter much whether or not he can tell you, or himself, what principles he is following.[23]

The distinctive structure of human practical activity in which an agent is motivated by her understanding of what is demanded by a role and by her self-identification as occupying a role, allows for a distinctive understanding of tools as tools. Many animals use natural objects to help them reach their organic ends. Some rational animals, such as chimps, even alter natural objects so as to make them more suitable for such use, and perhaps such animals can appropriately be said to use and make tools. But any agent that truly uses tools as tools must be motivated to use them in a particular way. For a tool is not only something that *can* be used in a certain way to attain a certain end but something that is *to be used* in a particular fashion, along with other equipment, to attain certain ends. Rocks and bricks, as well as hammers, can be used to hammer, but only *hammers* are instruments that are *supposed* to be used in hammering. A defective hammer is still a hammer, and it is *defective* only as a hammer, that is, a piece of equipment that is to be used in hammering.[24]

What is involved in being motivated to use a tool as a tool, that is, as something that is to be used in a certain fashion? First, being motivated to use an entity in a certain way because it is a tool of a certain sort is holistic in the sense that one doesn't intend to use a single tool type, but rather one acts in this way with this tool only if one also believes that there are other tools available that are to be used to accomplish further functions in the course of pursuing an end. What I do when I use a tool in a certain way because it is to be used in that way always involves assumptions regarding the status of other tool types that I take to be available and to be used in conjunction with this tool. This is to be used as *a hammer,* that is, as something to be used to hammer *nails.* These are to be used as *nails,* that is, things to be used to . . .

The normative structure of tools as tools, the fact that they are items that should be used in certain ways in certain situations along with other similarly defined tools, displays the structure that I have identified as "acting according to principle." When rational agents act because they are motivated by their belief that they are agents of a certain sort and that agents of that sort should act in this way in this situation, those agents treat *themselves* as standing under a norm that is independent from and higher than both what they are capable of doing and what they want to do. Similarly, to treat a tool as something that should be used in a certain way in a certain situation is both to treat the *object* as standing under the same type of abstract norm and to treat *oneself* as a certain kind of agent. The carpenter who uses a hammer as a hammer because it is to be used as a hammer, *infers* the use of this object from her belief that it is a hammer and her beliefs regarding how hammers are to be used. But in the same act the carpenter identifies herself as the kind of agent who is to use hammers in this way, as a carpenter, and infers how she is to act with this tool from this self-identification and from her understanding of how carpenters should use hammers. Because most human roles involve the use of tools, human action from principle thus typically involves both beliefs regarding how agents of a certain sort are to act and beliefs regarding how different sorts of equipment are to be used by agents of that sort.

The agent who uses a tool kit as tool kit displays a set of characteristic motivations. First, certain situations call for the use of the tool kit. These situations are defined in terms of both the situation in the external world and the situation in the intentional life of the group of rational animals who find the tools useful. When an instrumentally rational animal without fur finds that it is cold, it is appropriate that the animal come to want to become warmer. When that animal is a member of a group some of whose members are capable of building shelters, it is appropriate that the animal should come to believe that its companions can build a shelter that will satisfy its desire to become warmer and should come to desire that these companions do so. Since instrumentally rational animals mostly believe and desire what they should, this animal mostly comes to have these beliefs and desires. Now, carpenters are agents who, among other things, build shelters for members of their group who appropriately want them, and the carpenter who acts on principle will, in the circumstances described, be motivated to act as a carpenter to build a shelter. But carpenters are also agents who use certain tools in certain physical situations when they are building shelters, so our carpenter will start to build the shelter by using the tool appropriate to that situation. This use will in turn establish a new situation, which will call for the use of another tool, which will establish a new situation, which . . . will result in the shelter's being constructed. This result in turn will rationally motivate our original cold but instrumentally rational animal to move in.

The use of a tool kit thus involves a set of practices that prescribe how the various tools are to be used, practices that are specified both in terms of the preexisting physical and intentional environment and in terms of the environment constituted by the previous uses of items in the kit. There are certain "carpenter entry rules" (apologies to Sellars) that tell the principled carpenter when and how to set to work; there are certain rules that are internal to the practices of carpenters and that specify which tools are to be used in environments that are partially constituted by previous acts of carpentry; and there are "carpenter exit rules" that prescribe how the results of carpenter activity are to be used, by the carpenter and by others.

The level of normativity characteristic of tools as tools is thus distinct from the merely instrumental: the ability to be used to achieve an end is neither necessary nor sufficient to guarantee that some item counts as a tool. This distinctive level of equipmental normativity derives from the possibility of acting on principle according to norms appropriate to different social roles. Just as part of what it is to be a mother is to be motivated to act in certain situations by the fact that mothers act that way in those situations, part of what it is to be a carpenter is to be motivated to use specific objects in certain ways in certain situations to achieve certain ends because this is how carpenters use these objects in these situations to achieve those ends. To be a carpenter is, in part, to know how to use the carpenter's tool kit as carpenters use it to do what carpenters do and to act as one acts with these objects because this is how a carpenter *should* act. Tools, as tools, require practical agency.

Among the many instrumentally useful things that human beings can do in the context of flexible group activity, the most valuable is to communicate with one another. Perhaps most obviously, it is useful for human animals to communicate regarding the current state of the environment. Since human beings make a living by responding in an articulated, rapid, and socially differentiated fashion to a changing physical environment, there is a clear value in the ability of the members of a group to coordinate their activities in real time, on the fly. And, since different members of a group might be privy to different information at a given time, it is clearly valuable for the group members to be able to share information regarding the current state of the environment. Beyond this, however, the social instrumental utility of communication increases greatly when the members of a group become able to plan a common course of action ahead of time in response to a challenge. For this purpose, two other abilities beyond the ability to communicate regarding current conditions are crucial: the ability to communicate regarding past and possible future conditions and, most important, the ability to communicate one's own intentional states and about one another's intentional states.

Protohumans were highly instrumentally rational animals. In addition, they were animals who were in the process of developing the capacity to act on principle, the capacity to be motivated to perform an act because they believed that it was the right sort of act for an agent of their sort to perform. To predict, explain, and take into account the actions of agents who are motivated in these ways, it is supremely important for other agents to be able to discern the intentional states of their companions, because, as rational, these companions are motivated by these intentional states. Since protohumans were animals that survived through engaging in coordinated group activity, and since coordination depended on the ability of each member of the group to discern the intentional states of the other members, any means that helped to bring about that mutual understanding had important instrumental value. As it happens, human language as we know it is an excellent piece of equipment for communicating to each member of a group the intentional states of every other member of the group.

This communication of the contents of our intentional states takes place in virtually every act of assertion, as well as in other speech acts. Even a simple attempt to state a current fact about the world also contains information regarding the beliefs of the speaker. Therefore, if a language user can develop the trick of distinguishing the content of an assertion from the fact that it is being asserted by one of her companions, she can use this fact of an assertion's being asserted to help her make judgments regarding the beliefs and thus, inferentially, the desires of her companions. If I can recognize that you are asserting that there is a leopard over there by saying, "There is a leopard over here," then I can also recognize that you believe *that* there is a leopard over there. And if a protohuman were also equipped with the further ability in some way to monitor her own intentional states, then these two abilities could be combined to allow the individual agent to communicate her own intentional states *directly* to her companions, by saying that *she* wants to run away or that she believes there is a leopard over there, although she is not certain. Finally, any social group of primates that developed such abilities would have a considerable competitive advantage over any other such groups.

From this perspective, language, whatever else it might be or do, is an attractive tool kit for facilitating the transfer of information regarding the intentional states of the members of a group of instrumentally rational social animals. Language is a tool *kit* rather than a single tool because, as with all other equipmental structures, the end of language is achieved through the use of an interlinked set of tool types that are to be used with one another in prescribed ways. Hammers are to be used in certain situations in certain ways with nails to bring about the fastening of boards to one another, and the fastened boards are to be used in certain situations with posts, and so on. Similarly, different *sentences* are to be used in different situations, which uses in turn establish new situations in which other sentences are to be used and other actions performed. Just as there are carpenter entry and carpenter exit rules and rules internal to the practice of carpentry that taken all together specify how a carpenter's tools are to be used, so there are *language* entry and *language* exit rules and rules internal to the practice of speaking that taken all together specify how a speaker's tools—various sentences—are to be used.

The normative character of the principles that guide the practical activity of speaking is exhibited by the fact that the principles which define the use of the various sentences in a language specify how, and in what context, those sentences are to be used and what the appropriate use of one of these sentences in turn licenses. These principles are, in a word, *inference* rules that specify a sentence by prescribing the appropriate environmental and linguistic conditions for using the sentence and which other linguistic and overt acts can be appropriately performed if the sentence is itself appropriately used. For an instrumentally rational animal to come to be able to understand and use a language, then, is for that animal to come to act with these linguistic tools as a speaker of the language should act with these tools for the reasons that such speakers should act. That is, to learn a language, one must come to believe that one is an agent of a certain sort, come to believe that to be an agent of that sort is to *speak* in certain ways in certain situations, and finally come to infer that one should speak in certain ways in certain circumstances from these beliefs. An agent whose speech behavior is motivated in this way is a speaker of a language. As with any other type of teleological motivation, the principle of charity applies in the case of speech. All that is necessary for an agent to count as a speaker is, if she mostly acts appropriately given the rules, that she mostly performs the right inferences, not that she gets it right all of the time.

Speakers of a language are also in position to be interpreters of other speakers of the language. If one knows the principles that govern the linguistic behavior of an agent, one can infer from that behavior the beliefs and desires of that agent. One can do this because, as with all other forms of rational, teleological action, what a linguistic agent does is a function of what she believes and desires. Even rational agents that are motivated to act on principle do as they do because of their intentional states, even if the fact of their acting on principle adds a new range of intentional states that could motivate their behavior. But a speaker of a language is familiar with the principles that motivate all speakers of the language, insofar as they are acting as speakers of the language, to say the things they say when they are acting in a way that is appropriate for speakers of the language. So any speaker of a language is in an excellent position to be able to infer the intentional states of any other speaker of the language that are relevant to the explanation of what that other speaker says.

Speakers, however, do not have one set of intentional states that are relevant to the explanation of their linguistic behavior and an entirely different set of states that motivate their nonlinguistic behavior. All the intentional states of a rational agent hang together with all the rational behaviors of the agent in a single holistic teleological system. So any speaker of a language is also in an excellent position to interpret the intentional states of any other speaker of the language.

The advent of an articulate language not only allows for the possibility of a detailed understanding of the intentional states of a speaker of a language by other speakers of that language, but it also allows for the development of far more detailed contents of the intentional states of the speakers themselves. As Dennett points out, it is not only hard to imagine how a creature incapable of language could express a desire for "a two-egg mushroom omelette, some French bread and butter, and a half bottle of lightly chilled white Burgundy," it is also difficult to imagine how such an agent could come by such a desire.[25] Beliefs and desires have their content by virtue of their roles in the explanation of goal-directed behavior, and without language there is no possibility of any behavior's having such a goal, even when the behavior is that of an instrumentally rational agent. But the differentiation of tools in the linguistic tool kit allows us to develop a tool that has as its function the expression of the desire for a two-egg mushroom omelette. And because of this, language also allows us to distinguish between the desire for a two-egg mushroom omelette and the desire for what is on the plate over there, even when what is on the plate over there *is* such an omelette. For, after all, it is appropriate to *say* that one has the first desire in one intentional situation and to *say* that one has the second desire in another. So when a linguistic agent finds herself motivated to utter the first sentence, this is excellent evidence that she has the first desire, and when she is motivated to utter the second, this is excellent evidence that she has the second.

The possibility of a new, linguistic kind of rational behavior thus both demands and makes possible the development of a new, and far more detailed, type of intentional content. Because all intentional states get their contents from their roles in the explanation of different kinds of rational goal-directed action, the development of a new kind of principled, rational, goal-directed action with the capacity to make fine-grained distinctions among kinds of acts also makes possible—and makes necessary—a new kind of fine-grained content for intentional states.

As Davidson suggests, the fine-grained intensionality typical of human intentionality is thus integrally related to the ability to use and interpret language. One has such highly differentiated intentional states just in case one is capable of using and interpreting an articulate language. It is a necessary condition on being able to do this that an agent be capable of using tools as tools, for the ability to use a language is the ability to use a tool kit of a certain sort as a tool kit. It is a necessary condition on an agent's being able to use tools as tools that that agent is capable of acting on principle. For the ability to use a tool kit as a tool kit is a certain sort of ability to act on principle. But no agent can act in a practical way on principle unless she is also capable of acting in an instrumentally rational goal-directed fashion. The ability to act on principle in a practical way makes sense only as a manner in which an instrumentally rational animal makes its way in the world, and the contents of states that motivate such practical agents are intelligible only in terms of their place in a holistic, normatively constrained system of instrumentally rational behavior and explanation of that behavior. Finally, such rational behavior is possible only for an organic agent that acts in a goal-directed fashion for ends of its own and that is, as they say, an end in itself. Not all goal-directed agents are instrumentally rational, and not all instrumentally rational agents can act on principle. Only some agents that can act on principle in such a way as to use tools as tools create and use articulate language, but all linguistic agents with finely differentiated mental states are organisms that act for ends of their own.

The finely discriminated contents of human intentional states do depend on the existence of language, but this fact should never obscure the truth that language itself is possible only for a certain kind of instrumentally rational agent or the further truth that all instrumentally rational agents are, and must be, rational organisms that act for goals of their own. We are, truly, rational animals.

# NOTES

## PREFACE

1. Mark Okrent, *Heidegger's Pragmatism: Understanding, Being, and the Critique of Metaphysics* (Ithaca, N.Y.: Cornell University Press, 1988).

2. At that time I was very much gratified (as well as somewhat surprised) to find that the reception of my work by the analytic philosophers in my target audience was fair, open-minded, interested, and respectful. I suspect that this reception was partially a function of the fact that several important philosophers, including Hubert Dreyfus, Richard Rorty, John Haugeland, and Robert Brandom, were simultaneously developing many of these same parallels.

## INTRODUCTION

1. Donald R. Griffin, *Animal Thinking* (Cambridge, Mass.: Harvard University Press, 1984), 88–90.

2. Ibid., 37.

3. Fred I. Dretske, "The Intentionality of Cognitive States," in *Studies in Epistemology*, ed. Peter A. French, Theodore Edward Uehling, and Howard K. Wettstein, Midwest Studies in Philosophy 5 (Minneapolis: University of Minnesota Press, 1980), 284.

4. See Daniel C. Dennett, *Elbow Room: The Varieties of Free Will Worth Wanting* (Cambridge, Mass.: MIT Press, 1984).

5. Dean E. Wooldridge, *The Machinery of the Brain* (New York: McGraw-Hill, 1963), 82.

6. For now I am treating only instrumental rationality, although I don't hold that this is the only kind of rationality or the kind of rationality that is distinctive to language-using entities such as ourselves. I will argue that other forms of rationality piggyback on instrumental rationality, so we will need to understand this more basic type before attempting to comprehend its fancier cousins.

## CHAPTER 1

1. See Charles Taylor, *The Explanation of Behavior* (London: Routledge, 1964); Larry Wright, *Teleological Explanations: An Etiological Analysis of Goals and Functions* (Berkeley: University of California Press, 1976); Ruth Millikan, *Language, Thought, and Other Biological Categories: New Foundations for Realism* (Cambridge, Mass.: MIT Press, 1984).

2. Strictly speaking, of course, these are virtues that the pragmatic approach shares with many kinds of functionalism. As we shall see, the pragmatic approach has a distinct advantage over orthodox functionalist positions insofar as it has a much easier time dealing with the normativity of intentionality, the fact that by the very nature of the case intentional content can be normatively evaluated and that it is always possible that, for example, a belief can turn out to be false.

3. Ned Block, "Troubles with Functionalism," in *Perception and Cognition: Issues in the Foundations of Psychology*, ed. C. Wade Savage, Minnesota Studies in the Philosophy of Science 9 (Minneapolis: University of Minnesota Press, 1978), 261.

4. Ibid., 263.

## CHAPTER 2

1. See W. V. Quine, *Word and Object* (Cambridge, Mass.: MIT Press, 1960); Donald Davidson, *Inquiries into Truth and Interpretation* (Oxford: Clarendon Press, 1984).

2. See Davidson, "Thought and Talk," in *Inquiries into Truth and Interpretation*, 155–70.

3. For example, Donald Davidson, *Essays on Actions and Events* (Oxford: Clarendon Press, 1980).

4. Wright, *Teleological Explanations*, 39.

5. Throughout this work I assume a Davidsonian treatment of events rather than a Kimian one. For Davidson, events are particulars that are individuated by their causes and effects and that can be correctly described in a multitude of ways, as can any particular. For Kim, events are particulars that have the internal structure of being properties of objects at times. So in the Davidsonian view of events, the motion of the wasp wings and the preparation of the burrow are the same event described differently. For Kim, they are different events insofar as they involve different properties of the same or different objects. My preference for Davidson's view here is not really based upon any conclusive arguments. Rather it is based on the fact that the acceptance of Kim's view of events makes discussions of the

logic of goal-directed attributions exceedingly complex, convoluted, and difficult.

6. See Carl Hempel, "The Logic of Functional Analysis," in *Aspects of Scientific Explanation, and Other Essays in the Philosophy of Science* (New York: Free Press, 1965), 310. We will eventually see, in chapter 4, that goal-directed explanations do have a Hempelian structure. This structure, however, is incompatible with Wright's analysis of the content of goal-directed explanation.

7. Davidson, "Thought and Talk," 158.

8. Wright, *Teleological Explanations,* 39. Interestingly enough, this emphasis on success is also the cornerstone of the most influential competitor to Wright's position regarding the other main variety of teleological explanation, functional explanation. Robert Cummins holds that "if the function of something in a system *s* is to pump, then it must be capable of pumping in *s*." Cummins, "Functional Analysis," *Journal of Philosophy* 72, no. 20 (1975): 741–64. This, of course, is even stronger than the claim that if a thing has a function *f* in system *s,* it tends to do *f*. In any case, both of these assertions are clearly false, as we shall see.

Superficially, it might appear that Wright himself is committed to this same strong view regarding the relation of function and capacity. As Ruth Millikan's development of his position concerning functions makes clear, he is not. Wright is explicitly committed to the weaker tendency claim concerning goals, however.

9. There is a way of interpreting Wright's articulation of functional language, which we owe to Ruth Millikan, that involves an appeal to the evolutionary ancestors of the items that have a function. We will consider this articulation of Wright's work in a later chapter and see whether the same type of interpretation might be applied to goal-directed language. For now what is important is that Wright's clause (i), as stated, is simply and literally false for most goal-directed locutions.

10. Ultimately, I will use a variant of this strategy to unpack the explanatory force of goal-directed explanations. Needless to say, I will offer a noncircular articulation of (2) as well as an alternative to the Wrightian form of a potential clause (1).

11. Wright tries, feebly, to defend himself with "we don't understand causality." True, but we at least understand some of the logic of causal statements. More importantly, Ruth Millikan has suggested a quite plausible evolutionary understanding of this clause in Wright's corresponding formula for functional language that clears up much of the type/token confusion that is endemic to Wright's work and makes some sense of the causal character of Wright's articulation of teleological explanation. We will treat the analogue of this reconstruction for goal-directedness in chapter 4. For now, the important thing to note is that, as stated, Wright's proposal for understanding the explanatory force of goal-directed language is incoherent.

## CHAPTER 3

1. The main competitor to Wright's view of functions, the view developed by Robert Cummins, develops this fact into a decompositional analysis of functions. Roughly, the function of an item is the result that that item brings about that serves to further the process through which the system of which it is a part realizes some capacity. As we will see, the account of goal-directedness that is developed in this book is in some ways similar to Cummins's account of functions. The major problem with Cummins's account is that it implies that an item can't have a function that it does not, or cannot, bring about, and this seems clearly false.

2. Davidson has persuasively argued this point for the case of "that" clauses in "On Saying 'That,'" and the same arguments hold, *mutatis mutandis,* for clauses that attribute goals. See Davidson, "On Saying 'That,'" in *Inquiries into Truth and Interpretation* (Oxford: Clarendon Press, 1984), 93–108.

3. Jerry Fodor and Ernest Lepore, *Holism: A Shopper's Guide* (Oxford: Blackwell, 1992), 1–4.

4. It is of course possible that the only evidence available for the presence of a property is strongly holistic in this way but for the property itself not to be. If, for example, what it is for some event to have a goal were for it to be caused in the right way by a desire with a content equivalent to that goal, and if what it were for a desire to have a content involved some sort of representation relation, then the only evidence available for the attribution of a goal to a behavior would be strongly holistic, but, ontologically, the property of an event having a goal would not. So the above considerations are not conclusive against mentalism in teleology. Since, however, we are pursuing the strategy of making sense of intentionality by way of teleology, rather than the other way around, all we are interested in at this point is whether it is possible for one to make sense of goal attributions apart from intentional ones. So we are ruling appeals to mental intentionality out of court on strategic grounds. And, failing such an option, it is hard to see how the evidence in favor of goal attributions could be strongly holistic without the property itself also being strongly holistic.

5. Daniel Dennett, *Darwin's Dangerous Idea: Evolution and the Meanings of Life* (New York: Simon and Schuster, 1995), 68.

6. Richard Dawkins, *The Blind Watchmaker* (New York: W. W. Norton, 1986), 9.

7. Aristotle, *Metaphysics,* book IX, chapter 6, 1048b, lines 25–28, trans. W. D. Ross, in *The Basic Works of Aristotle,* ed. R. McKeon (New York: Random House, 1941), 826.

8. See Erwin Schrödinger, *What Is Life?* (Cambridge: Cambridge University Press, 1944), esp. 67–75.

9. It should be clear that the understanding of the living agent that is developed here is abstract relative to the chemistry of the organisms we are familiar with on earth. If there should be a silicon-based entity or a computer that preserves its unlikely structure by going through a program of interactions with its environment, then it counts as a living agent as much as wasps do.

## CHAPTER 4

1. Richard Feynman, *What Do You Care What Other People Think? Further Adventures of a Curious Character* (New York: Norton, 1988), 244.

2. Sellarsians, such as Robert Brandom, are fond of saying that things such as iron bars display reliable differential responses to their environments in the same sense that nonrational organisms do and that it is only entities capable of language that can be said to act in the logical space of reasons. This seems to me to be clearly false. Fran the wasp doesn't act because she recognizes the reasons for her acts, as we do; but there are reasons for her acts, while there are no reasons for the differential responses of iron bars. The mark of this distinction is that Fran will go into the burrow only when it is true that she is doing so in order to achieve the ultimate goal of reproducing.

3. Hempel, *Aspects of Scientific Explanation,* 310.

4. For example, Cummins makes this point in "Functional Analysis," 741–64.

5. See Davidson, "The Logical Form of Action Sentences" in *Essays on Actions and Events,* 118.

6. Ernst Mayr, *Evolution and the Diversity of Life: Selected Essays* (Cambridge, Mass.: Harvard University Press, 1975), 26.

7. See Elliott Sober, "Evolution, Population Thinking, and Essentialism," *Philosophy of Science* 47, no. 3 (1980): 350–83.

8. Ibid., 377–78.

9. Millikan, *Language, Thought, and Other Biological Categories.*

10. Ibid., 28.

11. Susan K. Mills and John H. Beatty, "The Propensity Interpretation of Fitness," *Philosophy of Science* 46, no. 2 (1979): 263–86.

12. Ibid., 277.

13. See John Maynard-Smith, "Optimization Theory in Evolution," *Annual Review of Ecology and Systematics* 9 (1978): 31–56.

14. Ibid., 52.

15. George C. Williams, *Adaptation and Natural Selection: A Critique of Some Current Evolutionary Thought* (Princeton, N.J.: Princeton University Press, 1966), 123.

## CHAPTER 5

1. Daniel Dennett, *Brainchildren* (Cambridge, Mass.: MIT Press, 1998), 327.

2. There is no doubt that this fact, combined with the fact that contexts in which goals are attributed are opaque, has led some philosophers, such as Dennett, to attribute beliefs and desires to viruses and even to thermostats. The argument seems to be: "An act has a goal just in case it can be explained by a corresponding desire. Thermostats and viruses do things that have goals. Therefore, they have desires that explain their goal-directed activities." We have seen, however, that the first premise of this argument is false. While it is true that an act has a goal *if* it is explained by a corresponding desire, it is not true that an act has a goal *only if* it is explained by such a desire. The class of actions in which desires explain goal-directed acts corresponds exactly with the class of *rational* actions.

3. If the condition persists in the wasps' environment, natural selection is often capable of generating a behavioral response in the offspring of surviving individuals and spreading it through the population, of course. But that is another matter.

4. Wooldridge, *Machinery of the Brain*, 82. Quoted in Dennett, *Elbow Room*, 11.

5. Griffin, *Animal Thinking*, 88–90.

6. If Wilfred Sellars is right, and I suspect he is in this respect, it is possible to have direct noninferential evidence regarding one's own intentional states. That, however, is another issue, one different from the question that is relevant here: whether an agent can have noninferential knowledge of the intentional states of another agent.

7. Griffin, *Animal Thinking*, 67.

## CHAPTER 6

1. See, for example, Daniel C. Dennett, "True Believers: The Intentional Strategy and Why It Works," in *The Intentional Stance* (Cambridge, Mass.: MIT Press, 1987), 13–35, and Davidson, "Thought and Talk," in *Inquiries into Truth and Interpretation*, 155–70. Revealingly, when Dennett introduces the intentional stance in this article, he explicitly helps himself to the notion of the ends of an agent without commenting on this assumption or giving any clue as to what he means by this notion. As I trust that the first four chapters of this book make clear, Dennett both needs this notion to make sense of his views on intentionality and, on his own grounds, is not entitled to it.

2. Several philosophers, most prominently Donald Davidson, have argued that the ability to respond to a mistake as a mistake is a necessary condition on an agent's being rational. See Donald Davidson, "The Emergence of Thought," in *Subjective, Intersubjective, Objective* (Oxford: Clarendon Press, 2001), 123–34. As the text should make clear, I entirely agree with this view. Many of these philosophers, again including Davidson, have gone on to argue that either linguistic competence or living in a certain kind of prelinguistic community is a necessary condition on responding to a mistake as a mistake. This seems to me to be false. As I argue in the text, any agent who is capable of self-correcting behavior and learning from her behavioral mistakes should count as responding to her mistakes as mistakes. Many nonlinguistic, nonsocial animals would seem to pass this test

3. While Davidson has been the *locus classicus* for the view that the entities that are the reasons for an action are also the entities that cause that action, there are grounds for believing that this reading of Davidson is based on a misinterpretation. See the excellent discussions in William Child, *Causality, Interpretation, and the Mind* (Oxford: Oxford University Press, 1994), and Frederick Stoutland, "Intentionalists and Davidson on Rational Explanation," in *Actions, Norms, Values: Discussions with Georg Henrik von Wright*, ed. Georg Meggle (Berlin: Walter de Gruyter, 1998), 191–208. Even if Davidson doesn't hold that the reasons for acts are identical with the causes of those acts, he does hold the weaker position that when a reason explains why an act occurs, it must be associated with an event that is the cause of that occurrence. He holds this view because of the considerations I discuss in the text.

4. The term "association" is used by Stoutland.

5. Davidson "Logical Form of Action Sentences," 105–22.

## CHAPTER 7

1. Davidson, "Thought and Talk," 163.

2. Ibid.

3. Notice that this assessment of the failure of the frog's act agrees with the assessment of the act that we could derive from an extension of Millikan's view on teleology but disagrees with the way in which the act would be assessed according to Wright's view. For Wright, the test for the goal of an act is to vary the circumstances in which the agent acts to see if it would still act in a way that tended to bring about a given result. In this case, that would involve varying the speed, direction, and other characteristics of the flying object to see whether the frog would still act so as to ingest it, which, of course, it would. So, in Wright's view, the goal of the tongue flicking

would be to eat lead. But in Millikan's view, the function of the tongue flick is fixed by the results of similar tongue flicks in the evolutionary past that aided and abetted frog reproduction—which, of course, was fly eating, not lead eating. So for Millikan, the act that results in eating lead is a failed attempt to eat flies. As the text makes clear, I have a strong intuition that Millikan is right in this assessment, although I come to it by a different route.

4. Even though in the case of the action of a rational agent this second description of the opacity would be preferable, I will continue to use the first. The reason for this is instructive. If the frog acted for a reason of its own, it would do so because it believed that this dark flying thing (which is a lead pellet) is a fly. So the opacity would arise out of this false belief. But frogs aren't rational and have no beliefs. So this isn't the source of the opacity in this case. The opacity arises out of the fact that the goal of the frog's act is that it be nourished. So it is only *as* an edible insect that there is a reason for the frog to catch this dark flying thing. Since the frog can catch a dark flying thing (a lead pellet) and fail to reach its goal, it is false to say that the goal of its act is to catch a dark flying thing even when that thing is a fly.

5. It is this level of opacity that infects contexts in which goals are specified, which surely explains why philosophers such as Dennett feel compelled to attribute reasons, beliefs, and desires to any system—including thermostats—that can be described as having a goal. If one recognizes, however, that (1) there is a principled difference between systems that can be assigned a goal extrinsically, such as thermostats, and living things that have their goals intrinsically and (2) the acts of nonrational agents can literally have goals, then Dennett's grounds for attributing beliefs and desires to apparently nonrational systems are undercut.

6. Alasdair C. MacIntyre, *Dependent Rational Animals: Why Human Beings Need the Virtues* (London: Duckworth, 1999), 37.

7. I owe to Heidegger the suggestion that one appropriate way to talk about opacity and intensionality is by speaking of what one intends an entity *as*. Heidegger was a student of Husserl, and Husserl had made the same distinction between sense and reference as Frege had made, generalizing this initially linguistic distinction to all intentional states. Heidegger, therefore, was well aware of the issues surrounding the opacity of intentional states, yet he agrees with Davidson in thinking that animals are incapable of intending anything as something.

8. Immanuel Kant, *Foundations of the Metaphysics of Morals*, trans. Lewis White Beck (Indianapolis: Bobbs-Merrill, 1959), 29.

9. Christine M. Korsgaard, *The Sources of Normativity* (New York: Cambridge University Press, 1996), 92.

10. Ibid., 93.

11. Kant's own formulations presuppose a simple identification of the ends of animal actions with the objects of their desires. If the views expressed

in this book are anywhere close to right, this is a mistake, as desire itself makes sense only in a context in which the ultimate ends of action are rooted in organic nature rather than desire. Korsgaard attempts to generalize Kant's problem to all nonprinciple determined ends, even when they are generated by the organic nature of the agent. How successful she is in this attempt remains unclear, at least to me.

12. Korsgaard, *Sources of Normativity*, 93.

13. Wilfrid Sellars, *Empiricism and the Philosophy of Mind* (Cambridge, Mass.: Harvard University Press, 1997), 73.

14. There are many fine books by contemporary primatologists which discuss these matters. Among the best are Frans de Waal, *Good Natured: The Origins of Right and Wrong in Humans and Other Animals* (Cambridge, Mass.: Harvard University Press, 1996), and Alison Jolly, *Lucy's Legacy: Sex and Intelligence in Human Evolution* (Cambridge, Mass.: Harvard University Press, 1999).

15. This fact was brought home to me in a graphic manner at a talk that Philip Kitcher gave at the Pacific Meetings of the American Philosophical Association several years ago. The meetings were held that year at the Claremont Hotel in Berkeley, and the talk was given in a room that provided a panoramic view of San Francisco Bay and the city beyond. Kitcher had just made the same point I am making here, that chimp communities are *much* smaller and less stable than human communities. At that point he turned around and gestured out the window, thereby indicating not only the community of the San Francisco Bay Area but also the United States, of which the Bay Area is but a small part. Parenthetically, I would like to thank Professor Kitcher for the germ of the idea that makes up the center of this chapter, that human life is distinguished from chimp life by the ability to act on principle, which I borrowed from this lecture.

16. In the talk I mentioned in the last note, Professor Kitcher alerted me to a standard interaction in chimp society that is often reported by primatologists. One chimp tentatively offers a banana to another, while apparently fighting with himself not to withdraw it. Meanwhile, the other chimp snatches the banana before the first can change his mind.

17. On the relation between individual and group selection, see Elliott Sober and David Sloan Wilson, *Unto Others: The Evolution and Psychology of Unselfish Behavior* (Cambridge, Mass.: Harvard University Press, 1998).

18. Kant, *Foundations of the Metaphysics of Morals*, 16–17.

19. This analysis of practicality is not entirely original to me. It is quite common in several different wings of the Kantian tradition. See Mark Okrent, "Heidegger and Korsgaard on Human Reflection," *Philosophical Topics* 27, no. 2 (1999): 47–76.

20. Any agent whose behavior can be predicted and explained by treating her as if she is motivated by what she infers to follow from the acceptance

of principle is, for practical purposes, rational. Given the social character of human intelligence, the ability to detect when an agent is acting on principle has enormous instrumental value. So our ability to predict and explain the actions of others by noticing the principles that they were instantiating—that is, our ability to detect cases of practical reasoning—probably developed in tandem with the ability to act in a practically rational manner. Also, given the fact that very few of us ever catch ourselves consciously inferring what we should do from general principles, it is a good bet that the reflective concept of action on principle probably arose out of our recognition that others were acting in such a way and only later came to be applied in the first person. As Sellars suggests in the case of our concept of a mental state, the concept of a practical inference can be applied in the first person only because it can be applied in the third.

21. See Martin Heidegger, *Being and Time,* trans., J. Macquarrie and E. Robinson (New York: Harper, 1962), division I, chapter 4.

22. Much of the material in this section and the last derives from my interpretation of the work of Martin Heidegger, who distinguishes two related forms of human goal-directed activity. According to Heidegger, we act in order to bring about certain determinate states of affairs; but when we do so, we act in a context provided by our self-understanding as agents of a certain sort that we understand ourselves to be instantiating. Although Heidegger recognizes the way in which these two forms of teleology are interdependent in human action, he gives a kind of formal priority to the self-interpreting action "for the sake of" realizing some possibility of human existence. This priority is related to his resistance to the traditional notion that we are rational *animals,* a position that I believe to be misguided. The ideas presented in these two sections are developed at much greater length and much greater detail in my "Korsgaard and Heidegger on Human Reflection," and "Intending the Intender: Or Why Heidegger Isn't Davidson," in *Heidegger, Authenticity, and Modernity,* vol. 1 of *Essays in Honor of Hubert Dreyfus,* ed. Mark Wrathall and Jeff Malpas (Cambridge, Mass.: MIT Press, 2000), 279–301. The material on the tool character of language is presented in greater detail in chapter 4 of my book *Heidegger's Pragmatism.*

23. Although this point is implicit in Heidegger, Hubert Dreyfus has articulated it most forcefully. Hubert Dreyfus, *Being-in-the-World* (Cambridge, Mass.: MIT Press, 1991).

24. My discussion of tools derives from Heidegger's analysis of equipment in *Being and Time.*

25. Dennett, "True Believers," in *Intentional Stance,* 20.

# BIBLIOGRAPHY


Aristotle. *Metaphysics*. Translated by W. D. Ross. In *The Basic Works of Aristotle*, edited by Richard McKeon. New York: Random House, 1941.

Bermúdez, José Luis. "Animal Reasoning and Proto-Logic." In *Rational Animals?* edited by Susan Hurley and Matthew Nudds, 127–37. Oxford: Oxford University Press, 2006.

———. *Thinking without Words*. Oxford: Oxford University Press, 2003.

Block, Ned. "Troubles with Functionalism." In *Perception and Cognition: Issues in the Foundations of Psychology*, edited by C. Wade Savage, 261–325. Minnesota Studies in the Philosophy of Science 9. Minneapolis: University of Minnesota Press, 1978.

Brandom, Robert. *Making It Explicit*. Cambridge, Mass.: Harvard University Press, 1994.

Child, William. *Causality, Interpretation, and the Mind*. Oxford: Oxford University Press, 1994.

Cummins, Robert. "Functional Analysis." *Journal of Philosophy* 72, no. 20 (1975): 741–65.

Davidson, Donald. *Essays on Actions and Events*. Oxford: Clarendon Press, 1980.

———. *Inquiries into Truth and Interpretation*. Oxford: Clarendon Press, 1984.

———. *Subjective, Intersubjective, Objective*. Oxford: Clarendon Press, 2001.

———. *Truth, Language and History*. Oxford: Clarendon Press, 2005.

Dawkins, Richard. *The Blind Watchmaker*. New York: W. W. Norton, 1986.

Dennett, Daniel C. *Brainchildren: Essays on Designing Minds*. Cambridge, Mass.: MIT Press, 1998.

———. *Darwin's Dangerous Idea: Evolution and the Meanings of Life*. New York: Simon and Schuster, 1995.

———. *Elbow Room: The Varieties of Free Will Worth Wanting*. Cambridge, Mass.: MIT Press, 1984.

———. *The Intentional Stance*. Cambridge, Mass.: MIT Press, 1987.

de Waal, Frans. *Good Natured: The Origins of Right and Wrong in Humans and Other Animals*. Cambridge, Mass.: Harvard University Press, 1996.

Dretske, Fred I. "If You Can't Make One, You Don't Know How It Works." In *Philosophical Naturalism*, edited by Peter A. French, Theodore E.

Uehling Jr., and Howard Wettstein, 468–82. Midwest Studies in Philosophy 19. Notre Dame, Ind.: Notre Dame University Press, 1994.

———. "The Intentionality of Cognitive States." In *Studies in Epistemology,* edited by Peter A. French, Theodore E. Uehling Jr., and Howard Wettstein, 281–94. Midwest Studies in Philosophy 5. Minneapolis: University of Minnesota Press, 1980.

Dreyfus, Hubert. *Being-in-the-World: A Commentary on Heidegger's* Being and Time, *Division I.* Cambridge, Mass.: MIT Press, 1991.

Feynman, Richard P. *What Do You Care What Other People Think? Further Adventures of a Curious Character.* New York: Norton, 1988.

Fodor, Jerry, and Ernest Lepore. *Holism: A Shopper's Guide.* Oxford: Blackwell, 1992.

Godfrey-Smith, Peter. *Complexity and the Function of Mind in Nature.* Cambridge: Cambridge University Press, 1996.

Griffin, Donald R. *Animal Thinking.* Cambridge, Mass.: Harvard University Press, 1984.

Haugeland, John. *Having Thought: Essays in the Metaphysics of Mind.* Cambridge, Mass.: Harvard University Press, 1998.

Heidegger, Martin. *Being and Time.* Translated by John Macquarrie and Edward Robinson. New York: Harper, 1962.

———. *The Basic Problems of Phenomenology.* Translated by Albert Hofstadter. Bloomington: Indiana University Press, 1975.

Hempel, Carl. *Aspects of Scientific Explanation, and Other Essays in the Philosophy of Science.* New York: Free Press, 1965.

Hurley, Susan, and Matthew Nudds, eds. *Rational Animals?* Oxford: Oxford University Press, 2006.

Jolly, Alison. *Lucy's Legacy: Sex and Intelligence in Human Evolution.* Cambridge, Mass.: Harvard University Press, 1999.

Kant, Immanuel. *Foundations of the Metaphysics of Morals.* Translated by Lewis White Beck. Indianapolis: Bobbs-Merrill, 1959.

Kim, Jaegwon. "On the Psycho-Physical Identity Theory." *American Philosophical Quarterly* 3, no. 3 (1966): 227–35.

Korsgaard, Christine M. *The Sources of Normativity.* New York: Cambridge University Press, 1996.

Kripke, Saul A. *Naming and Necessity.* Cambridge, Mass.: Harvard University Press, 1980.

Lewis, David. "Psychophysical and Theoretical Identifications." *Australasian Journal of Philosophy* 50, no. 3 (1972): 249–58.

MacIntyre, Alasdair C. *Dependent Rational Animals: Why Human Beings Need the Virtues.* London: Duckworth, 1999.

Maynard-Smith, John. "Optimization Theory in Evolution." *Annual Review of Ecology and Systematics* 9 (1978): 31–56.

Mayr, Ernst. *Evolution and the Diversity of Life: Selected Essays.* Cambridge, Mass.: Harvard University Press, 1975.

Merleau-Ponty, Maurice. *Phenomenology of Perception.* Translated by Colin Smith. London: Routledge, 1962.

Millikan, Ruth. "Biosemantics." *Journal of Philosophy* 86, no. 6 (1989): 281–97.

———. *Language: A Biological Model.* Oxford: Clarendon Press, 2005.

———. *Language, Thought, and Other Biological Categories: New Foundations for Realism.* Cambridge, Mass.: MIT Press, 1984.

———. "Styles of Rationality." In *Rational Animals?* edited by Susan Hurley and Matthew Nudds, 117–26. Oxford: Oxford University Press, 2006.

———. *Varieties of Meaning: The 2002 Jean Nicod Lectures.* Cambridge, Mass.: MIT Press, 2004.

Mills, Susan K., and John H. Beatty. "The Propensity Interpretation of Fitness." *Philosophy of Science* 46, no. 2 (1979): 263–86.

Okrent, Mark. "Heidegger and Korsgaard on Human Reflection." *Philosophical Topics* 27, no. 2 (1999): 47–76.

———. *Heidegger's Pragmatism: Understanding, Being, and the Critique of Metaphysics.* Ithaca, N.Y.: Cornell University Press, 1988.

———. "Intending the Intender: Or Why Heidegger Isn't Davidson." In *Heidegger, Authenticity, and Modernity,* edited by Mark Wrathall and Jeff Malpas, 279–301. Vol. 1 of *Essays in Honor of Hubert Dreyfus.* Cambridge, Mass.: MIT Press, 2000.

Proust, Jöelle. "Rationality and Metacognition in Non-Human Animals." In *Rational Animals?* edited by Susan Hurley and Matthew Nudds, 247–71. Oxford: Oxford University Press, 2006.

Putnam, Hilary. "The Meaning of Meaning." In *Language, Mind, and Knowledge,* edited by Keith Gunderson, 131–93. Minnesota Studies in the Philosophy of Science 7. Minneapolis: University of Minnesota Press, 1975.

———. "Psychological Predicates." In *Art, Mind, and Religion,* edited by W. H. Capitan and D. D. Merrill, 37–48. Pittsburgh: Pittsburgh University Press, 1973.

Quine, W. V. *Word and Object.* Cambridge, Mass.: MIT Press, 1960.

Schrödinger, Erwin. *What Is Life?* Cambridge: Cambridge University Press, 1944.

Sellars, Wilfrid. *Empiricism and the Philosophy of Mind.* Cambridge, Mass.: Harvard University Press, 1997.

Sober, Elliott, ed. *Conceptual Issues in Evolutionary Biology.* Cambridge, Mass.: MIT Press, 1994.

———. "Evolution, Population Thinking, and Essentialism." *Philosophy of Science* 47, no. 3 (1980): 350–83.

———, and David Sloan Wilson. *Unto Others: The Evolution and Psychology of Unselfish Behavior.* Cambridge, Mass.: Harvard University Press, 1998.

Sterelny, Kim. *Thought in a Hostile World: The Evolution of Human Cognition.* Oxford: Blackwell, 2003.

Stoutland, Frederick. "Intentionalists and Davidson on Rational Explanation." In *Actions, Norms, Values: Discussions with Georg Henrik von Wright,* edited by Georg Meggle, 191–208. Berlin: Walter de Gruyter, 1998.

Taylor, Charles. *The Explanation of Behavior.* London: Routledge, 1964.

Williams, George C. *Adaptation and Natural Selection: A Critique of Some Current Evolutionary Thought.* Princeton, N. J.: Princeton University Press, 1966.

Wooldridge, Dean E. *The Machinery of the Brain.* New York: McGraw-Hill, 1963.

Wrathall, Mark, and Jeff Malpas, eds. *Heidegger, Authenticity, and Modernity.* Vol. 1 of *Essays in Honor of Hubert Dreyfus.* Cambridge, Mass.: MIT Press, 2000.

Wright, Larry. *Teleological Explanations: An Etiological Analysis of Goals and Functions.* Berkeley: University of California Press, 1976.

Yablo, Stephen. "Mental Causation." *Philosophical Review* 101, no. 2 (1992): 245–80.

# INDEX

The letter *n* following a page number indicates a note on that page. It is followed by the note number

agents and agency: appropriate behavior and, 10–11; desires and, 62–63; goal attribution and, 64–66; goal-directedness and, 18–19, 30–31, 54, 60–74; intrinsic goals and, 65–66; living, teleology and, 60–74; misguided acts and, 56–57, 60–61, 104–5, 139–41, 145, 211n2; normative facts and, 127–33; rationality and, 10, 67, 135
appropriate behavior, 2, 4, 10–11, 42–44; definition, 5; logical behavioralism and, 6; scope of judgment of, 51–54; wasp *Sphex* and, 8–9; Wright and, 51–53
Aristotle, 35; "actualities," 70, 71; Darwin and, 68; typological versus population thinking and, 89–90
attitudes as dispositional states, 159–61

Beattie, John, 99–100
behavior patterns, preset. *See* stereotypical behavior
beliefs and desires, 108–9, 142, 166–67, 210n2; contents of, 125–26; counterfactual situations and, 131–33, 147–49; definitions, 161; goals and, 148–53; justification and, 138–39; Kant and, 190–91; new, coming to have, 143–45; normative constraints and, 129–33, 167; rational action and, 104–5; as reasons, 166–67; relevance and, 161–62
Brandom, Robert, 209n2

cats, shrews and, 175–76
causality, 31, 37, 39, 164; Davidson and, 211n3; internal state contents and, 171; rationality and, 154–61; reasons and, 153–56, 164; Wright and, 207n11

charity principle, 40, 54, 109; rationality and, 11, 146, 162, 166, 195; speech and, 202
chimpanzees, 184, 185–89, 197
communication, intentional states and, 200–204. *See also* language; speech
counterfactual situations: beliefs and, 131–33, 147–49; formula T and, 43, 56–57; goals and, 60; plovers and, 82; rationality and, 131–32, 147; success and, 124–25; wasp *Sphex* and, 54
covering laws, 36, 82–83, 84
Cummins, Robert, 102, 207n8, 208n1

Darwin, Charles: *On the Origin of Species*, 67–68
Darwinism: Millikan and, 91–103, 211n3; reproduction and, 91, 94–97, 99–102; typological versus population thinking, 89–103
Davidson, Donald, 28, 30, 41, 109, 136, 153, 208n1; causality and, 211n3; events and, 206–7n5; language and, 204; mistake response and, 211n2; speech and, 169–70
Dawkins, Richard, 69–70, 71, 74, 75–76
Dennett, Daniel, 68, 105–6, 112, 136, 171; beliefs and desires and, 210n2; goals, intrinsic versus attributed and, 212n5; intentional stance and, 107, 126, 210n1; language and, 203
Descartes, 22–23
desires and beliefs. *See* beliefs and desires
disjunction problem, 24–25
dispositional states, attitudes as, 159–61
dominance-hierarchical societies, 185–86

Dretske, Fred, 10–11, 22, 163; Griffin and, 6–7; versatile adaptability and, 4–7

environmental conditions, 7–8, 42–43, 123, 143; tracking of, 145–47. *See also* novelty
events, 206–7n5; acts and, 33–34; causality and, 154–55; covering laws and, 36; evaluability and, 25–26; intentionality and, 11, 13, 29–31; metaphysical theories of, 87–88; nonarbitrary norms and, 63–64, 67; normative evaluation of, 61–63; reasons and, 156–61; versus states, 149–50, 156; teleology and, 41–42; Wright and, 41–54
evolution. *See* Darwinism
existential generalizations, 157–58

Feynman, Richard, 75
field teleologist's problem, 28–31
fission-fusion groups, 184–85
fitness, reproductive success and, 99–102. *See also* reproduction
Fodor, Jerry, 22–23, 57, 163
formula T. *See under* Wright, Larry
Fran. *See* wasp sample (Fran)
frogs, intentionality and, 170–74, 211n3; opacity and, 212n4; reasons and, 166–67, 169–70
functionalism, 163, 171, 206n2; versus representationalism, 23–26

Galileo, 90
goal attributions, 28–33, 152–53, 208n4; beliefs and desires and, 142; contextualization and, 65–66; descriptive content of, 83–84; metaphor and, 41; plovers and, 56–57
goal-directedness, 30–40, 197; actual circumstances and, 46–47, 56–57; agency and, 30–31; appropriateness and, 42–44; causal explanation and, 37–39; covering law–style accounts of, 82–83; creation and, 67–68; description and, 32–35; explanation and, 39–40, 49, 207n10; language and, 32–37, 40–50; life stages and, 81–82; nomological explanation and, 36–37, 39, 73; normative evaluability and, 60–61, 67–68, 172; observation and, 30; prediction and, 37–38; rationality and, 39–41, 104; reasons and, 108–12; as strongly holistic property, 58–60; success and, 34–35; Wright and, 41–50
goals, 89–103; acts of agents and, 64–67; attributions, descriptive content of, 73–74, 79–80; desires and, 127; explanations and, 75–83; functional analysis and, 102–3; functions and, 83–89; as holistic properties, 141–42, 174, 208n4; nonarbitrary norms and, 73; predictability and, 98–99; species behavior and, 71–72, 122–25; types versus individuals and, 83–89
Griffin, Donald, 2–4, 11, 55–56, 78, 118; Dretske and, 6–7; versatile adaptability and, 2–3

Heidegger, Martin, 182–83, 194, 197, 212n7, 214n22
Hempel, Carl, 37, 80–82, 84–85, 88, 207n6
Heraclitus, 15, 61
hermeneutic circle problem, 59–60
holistic properties: definition, 57–58; goal-directedness and, 58–60; goals as, 174; natural numbers and, 57–58; reasons and, 132; strongly holistic properties, 54–60; systems and, 142–43, 164–65
human beings: communication and, 200–204; events and, 60–63; evolutionary story and, 183–89; practical rationality and, 183, 189–96; reason and, 168, 176–83; social groups and, 184–90

intentionality: attribution of, 13; definition, 12–13; evaluability and, 15; events and, 11, 13; existence and, 14, 16; goal-directedness and, 18; issues of, 12–17; minds and, 12; mysteries of, 13–17; normativity and, 14–15, 16, 165; opacity and, 169–73, 212n4; physicality of, 13–14, 16; puzzle of, 14; rationality and, 9–10, 111; response matrix, 26–27; second-order states and, 178–80, 190; speech and, 169–70; thought and action and, 17–18
intentionality, pragmatic approach, 11, 17–22; actions and, 20–21; beliefs and desires and, 164–65; functionalism and, 25–27; key features of, 27, 166; objections to, 167–68; versus representationalism, 25; summary, 163–65; virtues of, 21–22, 206n2

intentional stance versus teleological stance, 105–7, 126, 210n1

Kant, Immanuel, 130, 177–84, 189–96, 212n11; reasons and, 179–83
Kim, Jaegwon, 43
Korsgaard, Christine, 177–80
Kripke, Saul, 77

language, 183, 201–4, 205n6, 209n2. *See also* communication, intentional states and; speech
Lepore, Ernest, 57
life, 69–72, 209n9; beliefs and desires and, 105–8; empiricism and, 77–78; intrinsic goals and, 72–74, 105; modes of, 75–77; reproduction and, 72; Schrodinger and, 70–71; stages of, 81–82; structures and patterns, 71–72, 75–80, 166
linguist, field teleologist and, 28–31
logical behavioralism, 6, 159

MacIntyre, Alasdair, 175
Maynard-Smith, John, 100–101
Mayr, Ernst, 82–83, 89
Millikan, Ruth, 16, 19–20, 22, 163; failed acts and, 211n3; functions, evolution and, 92–99, 103
Mills, Susan, 99–100

natural selection, 67–68, 94–97, 99–102, 210n3. *See also* Darwinism
normative evaluability, 60–61, 67–68, 172
normativity, problem of, 178–80
norms, 61–63; evaluator and, 62; events and, 67; goals and, 73; instrumental rationality and, 127–33; intentions and, 15
novelty, 9, 20, 122–26; instrumental rationality and, 166, 176–77; plovers and, 120–21; practical rationality and, 193; rational adaptivity and, 104–5, 113, 125–27, 151–52; tits and, 135–36; wasp *Sphex* and, 113, 120–21

*On the Origin of Species* (Darwin), 67–68. *See also* Darwinism
ontological explanations, 16–17, 29, 55, 58–59, 159–61; goals and, 55, 58–59, 208n4
opacity: frogs and, 212n4; intentionality and, 169–73, 212n4

plovers, 1–7; actual circumstances and, 47–48; adaptive behavior, 2–4; consciousness and, 2; failure (error) and, 56–57; goal attribution and, 32–33; novelty and, 120–21; rationality and, 8–9, 20–21; relational goals and, 55–56; versus wasp *Sphex*, 118–22, 145
population thinking. *See* typological versus population thinking
practical rationality, 181–83, 185, 213n20; evolution of, 183–89; social roles and, 189–96
pragmatic approach to intentionality. *See* intentionality, pragmatic approach
Putnam, Hilary, 77

Quine, W. V., 28, 30, 109

radical interpreter, field teleologist and, 28–31
rationality: appearances of, 114–17; behavior motivation and, 153–61; causality and, 154–61; charity principle and, 11, 146, 162, 166, 195; human beings and, 168, 176–83, 189–96; intentional states and, 9–10, 111; Kant and, 130, 177–84; mistakes and, 21, 139–41, 211n2; practical, 181–83, 185, 189–96; success and, 128–29; teleology and, 104–5; types of, 111–12. *See also specific applications and issues*
reasons: attribution of, 132–33; beliefs and desires and, 104, 107–8, 136–41, 166–67; causality and, 153–56, 164; events and, 156–61; intrinsic goals and, 195–96; Kant and, 179–83; language and, 209n2; mental states and, 21–22; rational action and, 113–17, 120–22, 126, 129–30, 143–53, 157–61; teleology and, 135
representationalism, 22–26, 163
reproduction, 72; Darwinism and, 91, 94–97; fitness and, 99–102; natural selection and, 67–68; wasp sample (Fran) and, 80, 84; wasp *Sphex* and, 64–66

Schrödinger, Erwin, 70–71
S does B for the sake of G, 40–56, 60, 64, 83, 86–87; only if, 73–74, 79–84
Sellars, Wilfrid, 181, 209n2, 210n6

shrews, cats and, 175–76
Sober, Elliott, 90–91
speech, 169–70; charity principle and, 202. *See also* communication, intentional states and; language
*Sphex*. *See* wasp *Sphex*
stereotypical behavior, 123, 137, 173–74; tits and, 150–51; wasp *Sphex* and, 120–21, 134–35, 150–51
Stoutland, Frederick, 211n3
strongly holistic properties, 54–60. *See also* holistic properties

Taylor, Charles, 19–20
teleology: events and, 41–42; evolution and, 96–97; explanations and, 88–89, 134–36; language of, 19; living agents and, 60–74; locution and, 31; properties of, 19; rationality and, 104–5; reasons and, 135; teleological stance, 107; thought experiments and, 28–31; tits and, 135–36; typological versus population thinking and, 91–103. *See also* field teleologist's problem
thermostats, 105–7, 210n2
tits, 122, 125–28, 144, 146, 149–50, 167; beliefs and desires and, 135–42, 167; stereotypical behavior and, 150–51
tools: kits of, 199–204; normativity and, 199–200; roles and, 197–200
typological versus population thinking, 82–83, 89–92; Aristotle and, 89–90; Darwin and, 91; health versus disease and, 90–91

universal generalizations, 180–81

versatile adaptability, 3–4; Dretske, and, 4–7; plovers and, 2; rationality and, 10; tits and, 122; wasp *Sphex* and, 8. *See also* novelty

wasp sample (Fran), 79–90, 94, 97–98, 170, 209n2, 214n22
wasp *Sphex*, 7–9, 145; appropriate behavior, 8–9; burrow building, 31–33, 37, 42–43, 45, 64, 78; goal ascriptions and, 35–36; goal-directedness and, 20, 31, 33–34, 64; intentional mental states and, 10–11; nonrational behavior, 112–17; novelty and, 113, 120–21; versus plover, 118–22, 145; rationality and, 8–9, 20, 107, 110–11, 114–17; reproduction and, 64–66; stereotypical behavior, 150–51; teleology and, 8, 31–32; versatile adaptability and, 8
Williams, George, 101–2
Wooldridge, Dean, 7, 31, 78, 104, 114; actual circumstances and, 52
Wright, Larry, 19–20, 35; appropriate behavior and, 51–54; causality and, 50, 207n11; events and, 41–54; failed acts and, 211n3; formula T, 40–56, 60, 64, 83, 86–87; functions and, 208n1; mentalistic interpretation and, 44; quantifier scope and, 54; teleological explanation and, 207n8; tendency language and, 44–47, 207n9; type/token confusions and, 44–47